Partial Solutions

Introductory Chemistry and Fundamentals of Introductory Chemistry

Second Edition

Ebbing/Wentworth

David Bookin
MOUNT SAN JACINTO COLLEGE

Houghton Mifflin Company   Boston   New York

Senior Sponsoring Editor: Richard Stratton
Assistant Editor: Marianne Stepanian
Senior Manufacturing Editor: Priscilla J. Abreu
Marketing Manager: Penny Hoblyn

Copyright © 1998 by Houghton Mifflin Company. All rights reserved.

No part of this work may be reproduced or transmitted in any form or by any means, electronic or mechanical, including photocopying and recording, or by any information storage or retrieval system without the prior written permission of Houghton Mifflin Company unless such copying is expressly permitted by federal copyright law. Address inquiries to College Permissions, Houghton Mifflin Company, 222 Berkeley Street, Boston, MA 02116-3764.

Printed in the U.S.A.

ISBN: 0-395-87121-2

23456789- PR-01 00 99 98

CONTENTS

Preface	1
1. INTRODUCTION TO CHEMISTRY	3
Answers to Questions to Test Your Reading 3 / Answers to the Practice Exam 4	
2. MEASUREMENT IN CHEMISTRY	5
Solutions to Exercises 5 / Answers to Questions to Test Your Reading 7 / Solutions to Practice Problems 9 / Solutions to Additional Problems 11 / Solutions to Practice in Problem Analysis 13 / Answers to the Practice Exam 13	
3. MATTER AND ENERGY	14
Solutions to Exercises 14 / Answers to Questions to Test Your Reading 14 / Solutions to Practice Problems 17 / Solutions to Additional Problems 18 / Solutions to Practice in Problem Analysis 19 / Answers to the Practice Exam 19	
4. ATOMS, MOLECULES, AND IONS	20
Solutions to Exercises 20 / Answers to Questions to Test Your Reading 21 / Solutions to Practice Problems 23 / Solutions to Additional Problems 24 / Solutions to Practice in Problem Analysis 25 / Answers to the Practice Exam 25	
5. CHEMICAL FORMULAS AND NAMES	26
Solutions to Exercises 26 / Answers to Questions to Test Your Reading 27 / Solutions to Practice Problems 28 / Solutions to Additional Problems 29 / Solutions to Practice in Problem Analysis 30 / Answers to the Practice Exam 31	
6. CHEMICAL REACTIONS AND EQUATIONS	32
Solutions to Exercises 32 / Answers to Questions to Test Your Reading 38 / Solutions to Practice Problems 40 / Solutions to Additional Problems 43 / Solutions to Practice in Problem Analysis 44 / Answers to the Practice Exam 45	

7. CHEMICAL COMPOSITION 46
Solutions to Exercises 46 / Answers to Questions to Test Your Reading 51 /
Solutions to Practice Problems 53 / Solutions to Additional Problems 66 /
Solutions to Practice in Problem Analysis 70 / Answers to the Practice Exam 71

8. QUANTITIES IN CHEMICAL REACTIONS 72
Solutions to Exercises 72 / Answers to Questions to Test Your Reading 76 /
Solutions to Practice Problems 79 / Solutions to Additional Problems 86 /
Solutions to Practice in Problem Analysis 91 / Answers to the Practice Exam 91

9. ELECTRON STRUCTURE OF ATOMS 92
Solutions to Exercises 92 / Answers to Questions to Test Your Reading 93 /
Solutions to Practice Problems 95 / Solutions to Additional Problems 97 /
Solutions to Practice in Problem Analysis 98 / Answers to the Practice Exam 98

10. CHEMICAL BONDING 99
Solutions to Exercises 99 / Answers to Questions to Test Your Reading 104 /
Solutions to Practice Problems 108 / Solutions to Additional Problems 110 /
Solutions to Practice in Problem Analysis 112 / Answers to the Practice Exam 112

11. THE GASEOUS STATE 113
Solutions to Exercises 113 / Answers to Questions to Test Your Reading 119 /
Solutions to Practice Problems 121 / Solutions to Additional Problems 129 /
Solutions to Practice in Problem Analysis 130 / Answers to the Practice Exam 130

12. LIQUIDS, SOLIDS, AND ATTRACTIONS BETWEEN MOLECULES 131
Solutions to Exercises 131 / Answers to Questions to Test Your Reading 134 /
Solutions to Practice Problems 137 / Solutions to Additional Problems 141 /
Solutions to Practice in Problem Analysis 142 / Answers to the Practice Exam 142

13. SOLUTIONS 143
Solutions to Exercises 143 / Answers to Questions to Test Your Reading 147 /
Solutions to Practice Problems 152 / Solutions to Additional Problems 159 /
Solutions to Practice in Problem Analysis 163 / Answers to the Practice Exam 163

14. REACTION RATES AND CHEMICAL EQUILIBRIUM 164
Solutions to Exercises 164 / Answers to Questions to Test Your Reading 166 /
Solutions to Practice Problems 168 / Solutions to Additional Problems 171 /
Solutions to Practice in Problem Analysis 172 / Answers to the Practice Exam 173

15. ACIDS AND BASES — 174
Solutions to Exercises 174 / Answers to Questions to Test Your Reading 176 / Solutions to Practice Problems 180 / Solutions to Additional Problems 186 / Solutions to Practice in Problem Analysis 187 / Answers to the Practice Exam 187

16. OXIDATION-REDUCTION REACTIONS — 188
Solutions to Exercises 188 / Answers to Questions to Test Your Reading 194 / Solutions to Practice Problems 197 / Solutions to Additional Problems 202 / Solutions to Practice in Problem Analysis 205 / Answers to the Practice Exam 205

17. NUCLEAR CHEMISTRY — 206
Solutions to Exercises 206 / Answers to Questions to Test Your Reading 207 / Solutions to Practice Problems 211 / Solutions to Additional Problems 213 / Solutions to Practice in Problem Analysis 214 / Answers to the Practice Exam 214

18. ORGANIC CHEMISTRY — 215
Solutions to Exercises 215 / Answers to Questions to Test Your Reading 217 / Solutions to Practice Problems 223 / Solutions to Additional Problems 227 / Solutions to Practice in Problem Analysis 229 / Answers to the Practice Exam 229

19. BIOCHEMISTRY — 230
Solutions to Exercises 230 / Answers to Questions to Test Your Reading 231 / Solutions to Practice Problems 236 / Solutions to Additional Problems 239 / Solutions to Practice in Problem Analysis 240 / Answers to the Practice Exam 240

APPENDIX A — 241
Solutions to Exercises 241

APPENDIX B — 243
Solutions to Exercises 243

PREFACE

The Partial Solutions Manual provides worked-out answers to the following problems that appear in the second edition of *Introductory Chemistry* and *Fundamentals of Introductory Chemistry* by Ebbing and Wentworth. This includes detailed, step-by-step solutions for all in-chapter Exercises, all of the Questions to Test Your Reading, as well as the odd numbered Practice Problems and Additional Problems that appear at the end of the chapters. New to this edition are the Practice in Problem Analysis and Practice Exam questions, whose answers are also given.

Please note the following:

> **Significant figures:** The answer is first shown with 1 nonsignificant figure and no units, and the least significant digit is underlined. The answer is then rounded off to the correct number of significant figures and units are added.

D. B.

1. INTRODUCTION TO CHEMISTRY

■ Answers to Questions to Test Your Reading

1.1 Three examples, with properties given after each, are: 1) this book: solid, with a definite mass; 2) a pencil: solid, with a definite mass, and 3) air: a colorless gas, with an indefinite mass.

1.2 Three examples, with their descriptions given after each are: 1) wood burning in air: the flame is bright and produces smoke, gases, and ash; 2) baking soda and vinegar reacting: they form a clear solution and bubbles of carbon dioxide gas, and 3) copper and nitric acid reacting: they form a blue-green solution and a red-brown gas.

1.3 Two reactions and their energy changes are: 1) wood burning in oxygen to evolve heat energy and light energy, and 2) baking soda and vinegar reacting to evolve some heat energy and sound energy (fizzing of gas).

1.4 Chemistry is the science concerned with describing and explaining the different forms of matter and the chemical reactions (and accompanying energy changes) of matter.

1.5 Chemistry is also important in the following sciences. In biology, a major area of research is the detailed study of cell chemistry. Also, the understanding of many human diseases involves chemical reactions in the cell. In physics, the development of new materials, such as ceramic superconductors that exhibit superconductivity at temperatures closer to room temperature, depends on the chemical synthesis of the new material. In geology, the chemical reactions that occur within the earth or on its surface are studied to determine the conditions under which minerals and rocks form.

1.6 Some applications are: the chemical manufacture of colored pigments and dyes, obtaining perfumes from natural materials by extraction, and the manufacture of pottery and metals.

1.7 Copper was discovered when some blue-green rock (a copper ore) was heated with charcoal. Bronze was discovered when copper ore and tin ore were heated with charcoal.

1.8 Aristotle believed that all material things are made up of four primary or elementary substances: fire, air, water, and earth. He also believed that these substances differed in two properties: hot versus cold and dry versus wet. The primary substances are represented as the edges of a square, flanked by their associated properties on the corners of the square (see Figure 1.8).

1.9 Democritus and Leucippus believed that matter consisted of very small particles or atoms, which were too small to be seen with the naked eye, and which could not be cut into parts.

1.10 Lavoisier's experiments on combustion were characterized by using quantitative methods in his investigations. In particular, he used a balance as a tool to obtain the mass of the substances before and after burning.

1.11 Lavoisier proposed that the mercury combined with a component of the air, and that this component was identical to the gas that was formed when the original orange substance (mercury(II) oxide) was decomposed into liquid mercury and a gas.

1.12 The atomic theory of chemistry is associated with Dalton. This theory is based on the existence of matter in the form of atoms, and the idea that atoms of a given kind have a definite mass, or quantity of matter. Dalton was able to use his theory to explain the quantitative results that chemists obtained in their experiments.

1.13 a. An experiment is the observation of some natural phenomenon under controlled circumstances.

b. A law is a simple generalization from experiment.

c. A hypothesis is a tentative explanation of a law, or regularity of nature.

d. A theory is a tested explanation of some body of natural phenomena.

1.14 The scientific method advances a field of science by providing a general process for the testing of hypotheses. The scientist devises a series of experiments to test a hypothesis. As a result of these experiments, the scientist may amplify or modify the hypothesis. This scientist or others may then set up experiments to test this new hypothesis, and this hypothesis may be modified, and so forth. Eventually, a hypothesis that appears to agree with many experiments will be called a theory.

■ Answers to the Practice Exam

1. a   2. c   3. e   4. d   5. a   6. b   7. a   8. b   9. c   10. c

2. MEASUREMENT IN CHEMISTRY

■ Solutions to Exercises

Note on significant figures: Starting with Exercise 2.4, when the final answer is written for the first time, it is written with one nonsignificant digit, and the least significant digit is underlined. The final answer is then rounded to the correct number of significant figures.

2.1 In parts a, b, and e, the beginning zeroes are not significant, and do not appear in the answer. In parts b and c, the terminal zeroes are significant, and do appear in the answer.

   a. 6.9252×10^{-3} b. 6.300×10^{-2} c. 8.0200×10^{2}
   d. 9.002×10^{3} e. 7.07×10^{-1}

2.2 Move the decimal point the number of places indicated by the exponent. For a negative exponent, move the decimal to the left, and for a positive exponent, move it to the right. Add zeros where necessary.

   a. 0.0004834 b. 6250. c. 4.89

2.3 Beginning zeroes are not significant, and terminal zeroes are significant only where there is a decimal point.

   a. 4 sig. figs. b. 4 sig. figs. c. 4 sig. figs.
   d. at least 2 sig. figs. e. 5 sig. figs.

2.4 a. 97.0̲8 = 97.1 b. 28.2̲6 = 28.3
   c. 30.8̲2 = 30.8 d. 3.2̲1 = 3.2

2.5 a. 10.5 × (5.62 + 6.34) = 10.5 × 11.96 = 125.6 = 126
b. 5.84 − (1.23 ÷ 3.2) = 5.84 − 0.384 = 5.456 = 5.46

2.6 In the temperature conversion from absolute to Celsius, use this formula:

°C = K − 273
°C = 296 − 273 = 23°C

2.7 a. In the temperature conversion from Celsius to Fahrenheit, use this formula:

°F = 1.8°C + 32
°F = (1.8 × 35) + 32 = 63.0 + 32 = 95.0 = 95°F

b. In the temperature conversion from Fahrenheit to Celsius, use this formula:

°C = (°F − 32) ÷ 1.8
°C = (102 − 32) ÷ 1.8 = 70. ÷ 1.8 = 38.9 = 39°C

2.8 a. 2.58 g × (1 kg / 10^3 g) = 2.58 × 10^{-3} kg
b. 55.4 cm × (10^{-2} m / 1 cm) = 5.54 × 10^{-1} = 0.554 m
c. 7.11 L × (1 mL / 10^{-3} L) = 7.11 × 10^3 mL

2.9 a. 2.45 ns × (10^{-9} s / 1 ns) × (1 ms / 10^{-3} s) = 2.45 × 10^{-6} ms
b. 7.38 kg × (10^3 g / 1 kg) × (1 mg / 10^{-3} g) = 7.38 × 10^6 mg
c. 5.02 μL × (10^{-6} L / 1 μL) × (1 mL / 10^{-3} L) = 5.02 × 10^{-3} mL
d. 89.7 cm × (10^{-2} m / 1 cm) × (1 km / 10^3 m) = 8.97 × 10^{-4} km
e. 6.11 ms × (10^{-3} s / 1 ms) × (1 μs / 10^{-6} s) = 6.11 × 10^3 μs

2.10 The density is calculated as follows:

density = mass ÷ volume = (38.53 − 24.59 g) ÷ 4.50 cm³
= 3.097 = 3.10 g/cm³

MEASUREMENT IN CHEMISTRY ■ 7

2.11 Density = mass ÷ volume; rearranging gives mass = density x volume. (Note that ml = cm^3).

$$\frac{1.50 \text{ g}}{1 \text{ mL}} \times 8.4 \text{ mL} = 1 2.6 = 13 \text{ g}$$

2.12 Density = mass ÷ volume; rearranging gives volume = mass ÷ density, so

$$3.14 \times 10^4 \text{ g} \times \frac{1 \text{ cm}^3}{21.45 \text{ g}} = 1.464 \times 10^3 = 1.46 \times 10^3 \text{ cm}^3$$

■ Answers to Questions to Test Your Reading

2.1 Measurements consist of two parts, a measured number and a unit of measurement. When you measure the length of a rod, you compare its length to some standard measure of length, such as a ruler or a meter stick. Depending on the standard of measurement used, the length of the rod can be expressed in several ways. For example, when using a ruler, the length of a particular rod was 9.25 inches. Using a meter stick, the length of the same rod would be 0.235 meters, or 23.5 centimeters. The actual length of the rod is the same in both cases, although the measured number is different.

2.2 When the student pressed the equal sign, the calculator automatically converted the number to scientific notation. The zeros to the left of the 5 were dropped.

2.3. The answer is 10^{-4} or 0.0001.

2.4. The three base units are meter (length), kilogram (mass), and kelvin (temperature).

2.5 a. kilo   b. milli   c. centi   d. micro

2.6 There is one cubic decimeter and 1000 cubic centimeters in one liter.

2.7 The uncertainty in the measurement of the length of the room depends on the unit. If the tape measure is ruled in meters, then the uncertainty is ± 0.1 m. If the meter stick is ruled in centimeters, then the uncertainty is ± 0.1 cm. When measuring the room, the tape measure would probably be long enough so that the length of the room could be obtained in one measurement, but the meter stick would have to be moved repeatedly, and the length of the room would be obtained by adding up all of the measurements. Each time the stick is moved, an additional source of error is introduced, due to the inability to position the meter stick in the exact location, not in the measurement process. The measurements of two people won't agree exactly because each person's estimation of the last digit may differ. I would expect to be more accurate by measuring with the meter stick, which has a smaller uncertainty.

2.8 Writing 4.8300 g implies that the balance measures to ±0.0001 g, when the correct measurement implies that it only measures to ±0.001 g.

2.9 All five volume measurements agree on the first two digits, but there is disagreement on the third digit. Thus, three significant figures should be used in reporting the measured volume. The average is 41.2 mL, and the uncertainty is ±0.1 mL.

2.10 a. Counting is an example of an exact number.

b. A measured quantity is not an example of an exact number.

c. A definition like 3 ft = 1 yd is an example of an exact number.

d. A measured quantity is not an example of an exact number.

2.11 The third decimal place is unknown in both 66.45 and 61.35, so the third decimal place in 5.100 cannot be given. See the rule for addition and subtraction.

2.12 The "12" is a defined number and does not limit the number of significant figures in the answer. Thus the answer is 3.08̲3 or 3.08, with the same number of sig. figs. as 37.0 inches.

2.13 You should round only after the last step since you may lose some information or incur small errors from rounding up or down.

2.14 Yes, there is a difference. Heat is a form of energy, but temperature is a quantity you ascribe to a given point in a sample, and does not have to do with the total sample.

2.15 The freezing points are 0°C, 32°F, and 273 K.

2.16 This means that it is a scale that has zero for its lowest temperature.

2.17 The density of a substance is the mass of the substance per unit volume, usually expressed as g/cm^3 (or g/mL). The specific gravity of a substance is the ratio of the density of a substance to the density of water at 4°C. Specific gravity has no units.

2.18 Gypsum will float in methylene iodide because it has a smaller density and is therefore lighter, and will sink in carbon tetrachloride because it has a larger density.

2.19 The factor = g/grain = 1 g/15.43 grain = 0.06480̲8 = 0.06481 g/grain.

2.20 The factor = mg/μg = (1 mg/10^{-3} g) x (10^{-6} g/1 μg) = 10^{-3} mg/μg.

Solutions to Practice Problems

Note on significant figures: Starting with Problem 2.37, when the final answer is written for the first time, it is written first with one nonsignificant digit, and the least significant digit is underlined. The final answer is then rounded to the correct number of significant figures.

2.21 In converting a number in normal form to scientific notation, you should move the decimal point so as to obtain a number that is between 1 and less than 10. The number of places you move the decimal point gives the exponent. In parts b and d, note that the beginning zeroes are not significant, and do not appear in the answer. In part c, the terminal zeroes are significant, and do appear in the answer. The answers to each part are:

   a. 3.0402×10^4　　b. 6.58×10^{-3}　　c. 2.0900×10^2
   d. 3.5×10^{-6}　　e. 1.046×10^3

2.23 Move the decimal point to the right if the exponent is positive; to the left if negative.

   a. 890.45　　b. 0.005126　　c. 6.13
   d. 0.00007302　　e. 20,375.9

2.25　a. 26.8　　b. 0.004598　　c. 0.00783
   d. 8,536.　　e. 3,189.1

2.27　a. 3.69×10^{-3} m　　b. 3.489×10^4 m　　c. 9.2×10^{-11} s
   d. 5.67×10^{-8} g

2.29　a. 4.93 km　　b. 2.3 μs　　c. 5.68 mg
   d. 1.568 cm　　e. 9.3 ng

2.31 In counting the number of significant figures, remember that the beginning zeroes are not significant. Thus in part a the zero to the left of the decimal point is not significant. Terminal zeroes are significant only where there is a decimal point. Thus the terminal zeroes in parts b, c, and e are significant, but the terminal zero in part d is not significant:

   a. 4 sig. figs.　　b. 3 sig. figs.　　c. 4 sig. figs.
   d. at least 2 sig. figs.　　e. 2 sig. figs.

2.33 The 0.00800 of part c has 3 sig figs because the 3 beginning zeroes are not significant.

2.35 Round answers down when the first digit to be dropped is less than 5 and up when the it more than 5.

    a. 308.0   b. 45.15   c. 126.6
    d. 50.5092   e. 1.29379

2.37 a. 561.25 + 68.499 = 629.749 = 629.75
    b. 75.23 - 70.451 = 4.779 = 4.78
    c. 94.2 x 2.456 = 231.3 = 231
    d. 2.1 x 44.56 = 93.6 = 94
    e. 5.897 ÷ 7.8 = 0.756 = 0.76

2.39 a. (87.12 - 9.236) ÷ 5.489 = 14.189 = 14.19
    b. 65.456 x (23.698 - 20.613) = 201.93 = 201.9
    c. (583.21 - 502.13 + 10.23) ÷ 90.135 = 1.0130 = 1.013
    d. (73.54 ÷ 23.6) - 3.10 = 0.016 = 0.02
    e. (34.82 x 6.4 x 0.1439) - 2.68 = 29.4 = 29

2.41 a. 329 K   b. 49°C   c. 308 K   d. 185°C   e. 294 K

2.43   K = °C + 273
        30 + 273 = 303 K

2.45 No significant decimal places appear in any of the given temperatures, so the answers will not have any significant decimal places either. All digits to the left of the decimal point are significant in reporting temperatures.

    a. (85 - 32) ÷ 1.8 = 29.4 = 29°C
    b. (1.8 x -40) + 32 = -40.0 = -40.°F
    c. (-10 - 32) ÷ 1.8 = -23.3 = -23°C
    d. (1.8 x 55) + 32 = 131.0 = 131°F
    e. (1.8 x 125) + 32 = 257.0 = 257°F

MEASUREMENT IN CHEMISTRY ■ 11

2.47   °F = (°C x 1.8) + 32 = (-78 x 1.8) + 32 = -10_8_.4 = -108°F

2.49 a. 165 L x 1 mL/10^{-3} L = 1.65 x 10^5 mL
b. 48 ng x 10^{-9} g/1 ng = 4.8 x 10^{-8} g
c. 4.7 s x 1 ms/10^{-3} s = 4.7 x 10^3 ms

2.51 a. 7.53 in x $\dfrac{2.54 \text{ cm}}{1 \text{ in}}$ = 19._1_3 = 19.1 cm

b. 4.3 oz x $\dfrac{28.35 \text{ g}}{1 \text{ oz}}$ = 12_2_ = 1.2 x 10^2 g

2.53 a. 569 cm x (10^{-2} m/1 cm) x (1 km/10^3 m) = 5.69 x 10^{-3} km
b. 12.5 cm x (10^{-2} m/1 cm) x (1 dm/10^{-1} m) = 1.25 dm

2.55   0.77 Å x $\dfrac{100 \text{ pm}}{1 \text{ Å}}$ x $\dfrac{10^{-12} \text{ m}}{1 \text{ pm}}$ x $\dfrac{1 \text{ mm}}{10^{-3} \text{ m}}$ = 7._7_0 x 10^{-8} = 7.7 x 10^{-8} mm

2.57   Density = mass ÷ volume = 11.2 g ÷ 7.06 cm^3 = 1.5_8_6 = 1.59 g/cm^3

2.59   Density = mass ÷ volume = 0.975 g ÷ (5.97 - 5.60) mL = 2._6_3 = 2.6 g/cm^3

2.61   Density = mass ÷ volume; rearranging this definition to solve for mass gives:
mass = density x volume = 0.7908 g/mL x 24.48 mL = 19.3_5_8 = 19.36 g

2.63   Density = mass ÷ volume; rearranging this definition to solve for volume gives:
volume = mass ÷ density = 24.5 g ÷ 1.049 g/mL = 23._3_5 = 23.4 mL

■ Solutions to Additional Problems

2.65   All three volume measurements agree on the first three digits, but there is disagreement on the fourth digit. Thus, four significant figures should be used in reporting the measured volume. The average volume is 10.05 mL, and the uncertainty is ±0.01 mL.

2.67 Remember that terminal zeroes are significant when there is a decimal point.

    a. 9 sig. figs.   b. 3 sig. figs.   c. 2 sig. figs.
    d. 3 sig. figs.   e. 3 sig. figs.

2.69 a. 6.41   b. 15   c. 0.830
    d. 6.07   e. 247.9

2.71 2×10^{12} molecules

2.73 a. 4.37×10^{-5} b. 5.720×10^{5} c. 1.0×10^{-5}
    d. 1.5×10^{-5} e. 1.61×10^{-2}

2.75 a. 5.489 km   b. 723 ng   c. 21.64 km
    d. 6.50×10^{2} μL   e. 2.3 ms

2.77 $(1.8 \times 113°C) + 32 = 23\underline{5}.4 = 235°F$; $113 + 273 = 386$ K

2.79 $(93°F - 32) \div 1.8 = 3\underline{3}.8 = 34°C$; $34 + 273 = 307$ K

2.81 Volume = mass ÷ density = 0.0456 g ÷ 1.231 g/mL = $0.037\underline{0}4 = 3.70 \times 10^{-2}$ mL

2.83 Mass = density × volume = 1.063 g/cm^3 × 2.35 cm^3 = $2.4\underline{9}8 = 2.50$ g

2.85 Use a density equal to the specific gravity, and equate the volume to 1.50×10^3 mL:
Mass = density × volume = 1.28 g/mL × 1.50×10^3 mL = $1.9\underline{2}0 \times 10^3$ g, or 1.92 kg

2.87 0.5 mm × $\dfrac{10^{-3} \text{ m}}{1 \text{ mm}}$ × $\dfrac{1 \text{ μm}}{10^{-6} \text{ m}}$ = $\underline{5}.0 \times 10^2 = 5 \times 10^2$ μm

2.89 6 lb × 0.4536 kg/1 lb = $\underline{2}.72 = 3$ kg

2.91 60 μm × $\dfrac{10^{-6} \text{ m}}{1 \text{ μm}}$ × $\dfrac{1 \text{ cm}}{10^{-2} \text{ m}}$ × $\dfrac{1 \text{ in}}{2.54 \text{ cm}}$ = $2.3\underline{6} \times 10^{-3} = 2.4 \times 10^{-3}$ in

2.93 176 lb × 0.4536 kg/1 lb = $79.\underline{8}3 = 79.8$ kg

2.95 40 km/hr × 1 mi/1.609 km = $2\underline{4}.9 = 25$ mi/hr

Solutions to Practice in Problem Analysis

2.1 This problem involves the distance formula, distance = rate x time. The time to complete the walk is given by the distance divided by the rate. However, the distance is given in kilometers (6.0 km), and the rate in miles per hour (2.3 mi/hr). Therefore, you must convert the distance to miles using the appropriate conversion factor (1 mi = 1.609 km). The resulting answer for the time will be in hours. To finish the problem, you must convert the time into seconds using the appropriate conversion factor (1 hr = 3600 s).

2.2 This problem involves the density formula, density = mass ÷ volume. The mass of the liquid is given by the density multiplied by the volume. However, the density is given in grams per milliliter (0.82 g/mL), and the volume is given in quarts (2.2 qt). Therefore, you must convert the volume to milliliters using the appropriate conversion factors (1 qt = 0.9464 L, and 1 L = 1000 mL). The resulting answer for the mass will be in grams. To finish the problem, you must convert the mass to pounds, using the appropriate conversion factors (1 kg = 1000 g, and 1 lb = 0.4536 kg).

Answers to the Practice Exam

1. c   2. a   3. c   4. c   5. d   6. b   7. c   8. d   9. e   10. e

11. b   12. a   13. a   14. d   15. a

3. MATTER AND ENERGY

■ Solutions to Exercises

Note on significant figures: Starting with Exercise 3.3, when the final answer is written for the first time, it is written first with one nonsignificant digit, and the least significant digit is underlined. The final answer is then rounded to the correct number of significant figures.

3.1 Since 20°C is below the melting point of benzoic acid (122°C), it is in solid form.

3.2 Its physical properties are: colorless, physical state is a liquid, boiling point is 78°C, and freezing point is -114°C. Its only listed chemical property is that it burns, forming carbon dioxide and water.

3.3 Using the conversion factor of 1 cal/4.184 J:

48.6 J × 1 cal/4.184 J = 11.$\underline{6}$1 = 11.6 cal

3.4 Heat = 42 g × $\dfrac{0.129 \text{ J}}{1 \text{ g} \cdot °C}$ × (100°C - 26°C) = 4.$\underline{0}$1 × 10^2 = 4.0 × 10^2 J

3.5 sp heat = 5.68 J × $\dfrac{1}{0.875 \text{ g}}$ × $\dfrac{1}{(30.35°C - 25.47°C)}$ = 0.133$\underline{0}$ = 0.133 J/g·°C

■ Answers to Questions to Test Your Reading

3.1 Three examples are: 1) water in the liquid form is changed by freezing to ice, 2) water in liquid form is boiled and forms steam, and 3) water as ice is melted to form liquid water.

MATTER AND ENERGY ■ 15

3.2 Paper is a material that exists as a solid but not a gas or liquid.

3.3 The name is the melting point (it usually equals the freezing point).

3.4 The term vapor describes the gaseous state of a substance that normally exists as a liquid or solid.

3.5 As you add energy to a solid substance, you raise its temperature and it melts. On a molecular level, while in the solid phase, the molecules of the substance are in a fixed position. The energy that has been added causes the molecules of the substance to begin moving away from their fixed positions, and the distance between the molecules increases. When the temperature reaches the melting point, the substance becomes a liquid. As you continue to add energy to the liquid, the temperature continues to rise. The molecules are moving faster and faster, and the distance between the molecules continues to increase. When sufficient energy has been added so that the temperature reaches the boiling point, the substance begins to boil. By now, the molecules are very far apart, and the substance is a gas.

3.6 As you remove energy from a liquid substance, you lower its temperature and it freezes. On a molecular level, while in the liquid phase, the molecules of the substance are moving in continuous random motion. The energy that has been removed causes the molecules of the substance to slow down and begin moving closer to each other. When the temperature reaches the freezing point, the molecules take up fixed positions, and the substance becomes a solid. The condensation of a gas is similar. As you remove energy from the gas, the molecules begin to slow down and move closer together. When sufficient energy has been removed so that the temperature reaches the boiling point, the substance begins to condense. By now, the molecules are much closer together, and the substance is a liquid.

3.7 Three examples are: 1) sawing a log into boards, 2) changing liquid water into ice or steam, and 3) dissolving solid salt in water.

3.8 Three examples are: 1) burning a log to form smoke, gaseous products, and ashes, 2) reaction of iron with oxygen to form iron oxide (rust), and 3) reacting mercury metal with oxygen to form mercury(II) oxide.

3.9 Three examples are: 1) melting point and boiling point, 2) density, and 3) color.

3.10 Iron reacts with oxygen to produce a red-brown iron oxide (rust), and bromine, a red-brown liquid, reacts with sodium to form a white solid.

3.11 A substance is a material that cannot be separated into different materials by physical processes. An example is carbon dioxide.

3.12 A mixture is a material that can be separated by physical processes into two or more substances. An example is a mixture of salt and water.

3.13 Filtration depends on pouring a solution with a suspended precipitate through a filter cone. The clear liquid passes through the filter, leaving the precipitate behind on the paper.

3.14 When a solution of salt in water is heated to the boiling point, the water vapor passes out of the distillation flask, into the condenser where it cools and runs out into the receiver flask. At the end of the distillation, pure salt remains in the distillation flask, and pure water has collected in the receiver flask.

3.15 Two examples are: 1) a liquid solution of sodium chloride in water, and 2) air, which is a gaseous mixture of nitrogen, oxygen, and several other substances.

3.16 Two examples are: 1) a mixture of sugar and salt crystals, and 2) beach sand, which is a mixture of calcite, silica, and other materials.

3.17 An element is a substance that cannot be decomposed by any chemical reaction into simpler substances. Three examples are: 1) mercury metal, 2) oxygen gas, and 3) iron metal.

3.18 A compound is a substance composed of two or more elements chemically combined. Three examples are: 1) sugar, 2) salt, and 3) aspirin.

3.19 A pure compound, whatever its source, always contains definite or constant proportions of the elements, by mass.

3.20 Energy is the potential or capacity to move matter.

3.21 Kinetic energy is the energy of matter in motion. Potential energy is matter with a potential for motion. As an example, if you hold a flashlight battery in the air, it has potential energy because it has the potential for motion. If you let it drop, its potential energy is converted to kinetic energy.

3.22 Four other forms of energy mentioned in the chapter are chemical energy, light energy, food energy, and nuclear energy.

3.23 The specific heat of a substance is the quantity of heat required to raise the temperature of one gram of that substance by one degree Celsius. In SI units, the specific heat of water is 4.18 J/(g·°C).

3.24 Einstein's theory of relativity contains a mathematical relationship between mass and energy, and therefore, mass can be regarded as just another form of energy. The law of conservation of energy is a more general statement than Lavosier's law of conservation of mass, and include it as a specific type of energy conservation. Thus, Lavosier's law is still correct. Also, in normal chemical processes, the amount of mass that is converted to energy in the form of heat is so small that it can be ignored. The law of conservation of mass is therefore more useful in everyday work.

Solutions to Practice Problems

Note on significant figures: Starting with Problem 35, when the final answer is written for the first time, it is written first with one nonsignificant digit, and the least significant digit is underlined. The final answer is then rounded to the correct number of significant figures.

3.25 It is a liquid.

3.27 It is a gas.

3.29 Its physical properties are: colorless, 69°C boiling point, and -95°C freezing point. Its chemical property is that it burns to form CO_2 and H_2O.

3.31 Its physical properties are: colorless, oily liquid, and density of 1.59 g/cm^3. Its chemical properties are: dilates blood vessels, and explodes forming N_2, CO_2, H_2O, and O_2.

3.33 a. mixture b. element c. compound d. mixture

3.35 Using 121 kJ = 1.21×10^5 J, and the conversion factor of 1 cal/4.184 J:

1.21×10^5 J x 1 cal/4.184 J = $2.8\underline{9}1 \times 10^4$ = 2.89×10^4 cal

3.37 Using the conversion factor of 4.184 J/1 cal:

12.0×10^3 cal x 4.184 J/1 cal = $5.0\underline{2}1 \times 10^4$ = 5.02×10^4 J

3.39 Heat = 235 g x $\dfrac{0.901 \text{ J}}{1 \text{ g} \cdot \text{°C}}$ x (15.0°C) = $3.1\underline{7}6 \times 10^3$ = 3.18×10^3 J

3.41 Heat = 65.4 g x $\dfrac{0.741 \text{ J}}{1 \text{ g} \cdot \text{°C}}$ x (38.5 - 20.0)°C = $8.9\underline{6}5 \times 10^2$ = 897 J

3.43 sp heat = 151.5 J x $\dfrac{1}{18.9 \text{ g}}$ x $\dfrac{1}{(39.9°C - 23.2°C)}$ = $0.47\underline{9}9$ = 0.480 J/g·°C

3.45 Increase, °C = 1071 J x $\dfrac{1}{36.0 \text{ g}}$ x $\dfrac{1 \text{ g} \cdot \text{°C}}{2.43 \text{ J}}$ = $12.\underline{2}4$ = 12.2°C

3.47 Increase, °C = 164.8 J × $\dfrac{1}{15.5 \text{ g}}$ × $\dfrac{1 \text{ g} \cdot °C}{0.449 \text{ J}}$ = 23.67°C

Final temp = 21.5°C + 23.67°C = 45.17 = 45.2°C

■ Solutions to Additional Problems

3.49 a. Gas   b. Gas   c. Gas   d. Liquid   e. Solid

3.51 a. Physical   b. Chemical   c. Physical   d. Physical   e. Chemical

3.53 a. Solid   b. Solid   c. Solid   d. Solid   e. Gas

3.55 a. Solid   b. Liquid   c. Liquid   d. Gas   e. Gas

3.57 a. Substance   b. Het. mixt.   c. Solution   d. Substance

3.59 a. Physical   b. Chemical   c. Physical   d. Chemical

3.61 453 Btu × 252 cal/Btu = 1.141 × 10^5 = 1.14 × 10^5 cal

453 Btu × 252 cal/Btu × 4.184 J/cal = 4.776 × 10^5 = 4.78 × 10^5 J

3.63 3.5 × 10^4 cal × $\dfrac{1 \text{ g} \cdot °C}{1 \text{ cal}}$ × $\dfrac{1}{(35.0 - 20.0)°C}$ = 2.33 × 10^3 = 2.3 × 10^3 g

3.65 $\dfrac{4.18 \text{ J}}{1 \text{ g} \cdot °C}$ × 5.42 g × (45.1 - 20.8)°C = 550.5 = 551 J

Solutions to Practice in Problem Analysis

3.1 This problem is solved by first calculating the heat that was lost by the silver bar. This is then set equal to the heat gained by the water. The heat lost by the silver bar is equal to the mass of the bar (2563 g) multiplied by the specific heat of silver (0.449 J/(g•°C)), and multiplied by the temperature change of the bar (85°C - 36°C = 49°C). This quantity of heat must equal the mass of the water (966 mL, or 966 g) multiplied by the specific heat of water (4.18 J/(g•°C)), and multiplied by the temperature change for water (ΔT). Only the temperature change for water is unknown. To solve for the temperature change, divide both sides of the equation by the mass of water and the specific heat of water.

3.2 The caloric content of the cereal can be determined by calculating the caloric content of the protein, carbohydrates, and fat individually, and then adding up. The mass of the protein (8 g) is multiplied by the heat capacity for protein (17 kJ/g), the mass of the carbohydrates (26 g) multiplied by the heat capacity for carbohydrates (17 kJ/g), and the mass of the fat (2 g) is multiplied by the heat capacity for fat (38 kJ/g). The resulting sum is given in kJ, and will have to be converted to Calories (1 cal = 4.184 J, or 1 Cal = 4.184 kJ).

Answers to the Practice Exam

1. b   2. c   3. b   4. a   5. d   6. a   7. d   8. d   9. c   10. c

11. b   12. b   13. a   14. d   15. e

4. ATOMS, MOLECULES, AND IONS

■ Solutions to Exercises

Note on significant figures: Starting with Exercise 4.4, when the final answer is written for the first time, it is written first with one nonsignificant digit, and the least significant digit is underlined. The final answer is then rounded to the correct number of significant figures.

4.1   The atomic number of the atom is 17 (number of protons). The mass number is 17 + 18 = 35. From the list of elements on the inside back cover, we see that the symbol of the element with atomic number 17 is chlorine. The isotope symbol is:

$$^{35}_{17}Cl$$

4.2   The left subscript of 9 gives the number of protons for fluorine. The left superscript gives the mass number, so the number of neutrons in the nucleus is

   number of neutrons = mass number - atomic number

   number of neutrons = 19 - 9 = 10

4.3   From the inside back cover, we see that boron has an atomic number of 5, indicating 5 protons and 5 electrons. The number of neutrons = mass number - atomic number; so 11 - 5 = 6 neutrons. The protons and neutrons are located in the nucleus whereas the electrons are located around and far away from the nucleus.

4.4   To find the atomic weight of copper, multiply each isotopic mass by its fractional abundance. Then add the products to obtain the atomic weight:

   62.94 amu x 0.6917 = 43.5$\underline{3}$6 amu
   64.93 amu x 0.3083 = 20.0$\underline{1}$8 amu

   atomic weight   = 63.5$\underline{5}$4 =   63.55 amu

ATOMS, MOLECULES, AND IONS   21

4.5   The element in Group IIA, period 3 is magnesium. The symbol of the element is Mg.

4.6   Using the periodic table (Figure 4.9), we see that nitrogen (N) and phosphorus (P) are the nonmetallic elements in Group VA.

4.7   Methane is composed of carbon and hydrogen, both nonmetals, indicating that methane is molecular. Because it is a gas at room temperature, its melting point is below room temperature, which is expected of a molecular substance.

4.8   This molecule consists of phosphorus and oxygen so its molecular formula will contain their respective symbols P and O. Since there are four atoms of phosphorus and ten oxygen atoms, the formula is P_4O_{10}.

■ Answers to Questions to Test Your Reading

4.1   According to Dalton, an element is a type of matter composed of only one kind of atom, which always has certain specific properties. He defined a compound as a type of matter composed of atoms of two or more elements chemically combined in fixed proportions. He explained a chemical reaction as a rearrangement of the atoms present in the reacting substance, to give new chemical combinations in the substances formed by the reaction.

4.2   Atomic theory states that because the number of each kind of atom is the same before and after a reaction, the total mass remains constant. Atomic theory says that because a compound contains atoms in definite, fixed proportions, and because atoms have define masses, compounds must be present in definite proportion by mass.

4.3   Cathode rays are produced in an electric discharge tube containing two electrodes, a cathode and an anode. When high voltage is fed to the electrodes, a beam of radiation is emitted from the cathode, called a cathode ray. The particles that make up the ray are electrons, which are attracted to the positive anode.

4.4   The results of the experiment showed that most alpha particles passed undeflected through the metal foil, but a few were deflected nearly backward. Rutherford concluded that atoms in the foils had a nucleus, or a positively charged atomic core that takes up very little space in the atom but has most of its mass.

4.5   The one- or two-letter symbol identifies the element. The left subscript gives the atomic number as well as the number of protons and electrons. The left superscript gives the atomic mass or total of the number of protons and neutrons. The number of neutrons is obtained by subtracting the subscript from the superscript.

4.6   Inside the sodium nucleus are 11 protons and 12 neutrons; outside the sodium nucleus are 11 electrons.

4.7 The isotopes of the elements were thoroughly mixed during the formation of earth, and chemical and most physical processes do not significantly alter this mixture.

4.8 The carbon-12 isotope is the modern standard used for obtaining the masses of other atoms and atomic weights.

4.9 Modern periodic tables differ in the respect that they are arranged by increasing atomic numbers, not the atomic weights used by Mendeleev.

4.10 The letter "B" indicates that copper is a transition-metal element.

4.11 The Group IA metals, which include lithium, sodium, and potassium, are similar in that they are all soft metals that react readily with water. Their physical properties also change in a regular fashion.

4.12 The two rows at the bottom are not connected because it is difficult to include a row of 32 elements, with all the usual information, on a text page.

4.13 A metalloid, like silicon, has properties intermediate between those of a metal and those of a nonmetal. It is also a semiconductor of electricity.

4.14 The molecular basis of a substance refers to the actual building blocks of elements and compounds. The term molecular is used loosely because molecules are only one kind of building block; other kinds are atoms and ions.

4.15 A molecule of an element consists of only one kind of atom. For example, the nonmetals, hydrogen and oxygen, occur in nature as diatomic (two-atom) molecules.

4.16 The two main types of compounds are molecular compounds (example: water) and ionic compounds (example: sodium chloride). Ionic compounds tend to be solids, and melt at high temperatures, whereas many molecular compounds are liquids or gases, and have relatively low melting points.

4.17 A molecular formula is a notation that uses atomic symbols with subscripts to convey the exact number of atoms of different elements in a molecule of the substance. As an example, the molecular formula of water is H_2O.

4.18 A structural formula is a chemical formula that shows what atoms are bonded to one another. As an example, the structural formula of water is H-O-H.

4.19 The molecular formulas are: hydrogen, H_2; nitrogen, N_2; oxygen, O_2; chlorine, Cl_2; and sulfur, S_8. For carbon, the formula would be C.

ATOMS, MOLECULES, AND IONS   23

4.20 Sodium chloride has a structure with equal numbers of sodium ions, Na^+, and chloride ions, Cl^-. The strong attraction between positive and negative charges holds the ions together to form a crystal of sodium chloride. The total number of ions in the crystal varies, but there is an extremely large number of sodium ions and chloride ions, and because they are in equal numbers the crystal is electrically neutral. The ratio of the number of sodium ions to chloride ions is one to one. Each Na^+ ion is surrounded by 6 Cl^- ions.

4.21 Since $CaCl_2$ contains a metal and a nonmetal, it should be an ionic compound. Therefore, a water solution of $CaCl_2$ should conduct electricity.

4.22 Table sugar is a molecular substance. Therefore a water solution of sugar should not conduct electricity.

■ Solutions to Practice Problems

Note on significant figures: Starting with Problem 4.35, when the final answer is written for the first time, it is written first with one nonsignificant digit, and the least significant digit is underlined. The final answer is then rounded to the correct number of significant figures.

4.23 a. $^{20}_{10}Ne$ b. $^{28}_{14}Si$ c. $^{31}_{15}P$ d. $^{39}_{19}K$ e. $^{55}_{25}Mn$

4.25 a. 7p + 7n   b. 15p + 17n   c. 16p + 18n   d. 17p + 18n   e. 24p + 28n

4.27 Zinc-64 has 30 protons, 34 neutrons, and 30 electrons. The protons and neutrons are in the nucleus; the electrons are far outside the nucleus.

4.29 Selenium-80 has 34 electrons.

4.31 Arsenic-75 has 33 protons and 42 neutrons in the nucleus and 33 electrons outside the nucleus.

4.33 Nitrogen-14 is far more abundant.

4.35 To find the atomic weight of boron, multiply each isotopic mass by its fractional abundance. Then add the products to obtain the atomic weight:

$$10.01 \text{ amu} \times 0.197 = 1.9\underline{7}2 \text{ amu}$$
$$11.01 \text{ amu} \times 0.803 = 8.8\underline{4}1 \text{ amu}$$

atomic weight = $10.8\underline{1}3$ = 10.81 amu

4.37 To find the atomic weight of the unknown element, multiply each isotopic mass by its fractional abundance. Then add the products to obtain the atomic weight:

$$19.99 \text{ amu} \times 0.9092 = 18.175 \text{ amu}$$
$$20.99 \text{ amu} \times 0.00257 = 0.05394 \text{ amu}$$
$$21.99 \text{ amu} \times 0.088 = 1.939 \text{ amu}$$

atomic weight = 20.168 = 20.17 amu (The element is neon.)

4.39 a. In: indium b. Si: silicon c. Mn: manganese
  d. Ti: titanium e. As: arsenic

4.41 Bismuth(Bi) is the only metal in Group VA.

4.43 Phosphorus: atomic no. = 15; atomic wt. = 30.97, group VA, period 3, nonmetal.

4.45 a. C + O form molecular compound(s)
  b. N + O form molecular compound(s)
  c. Li + O form ionic compound(s)
  d. Ca + Cl form ionic compound(s)
  e. S + Cl form molecular compound(s)

4.47 Sodium, sulfur, and oxygen should form an ionic compound of the type $Na_xS_yO_z$, where Na^+ is the cation and the $S_yO_z^{n-}$ is a polyatomic anion. This prediction agrees with the high melting point.

4.49 a. H_2SO_4 b. Cl_2O_7 c. N_2O_4
  d. C_3H_8 e. N_2F_4

Solutions to Additional Problems

4.51 Sodium-24 has 11 protons, 11 electrons, and 13 neutrons.

4.53 The isotope symbol for this calcium isotope is:

$$^{47}_{20}Ca$$

ATOMS, MOLECULES, AND IONS   25

4.55 They differ by two neutrons; they are alike in that both have 28 protons and 28 electrons.

4.57 The proton is 1.8×10^3 times heavier than the electron.

4.59 To find the atomic weight of chromium, multiply each isotopic mass by its fractional abundance. Then add the products to obtain the atomic weight:

$$
\begin{array}{rcl}
49.95 \text{ amu} \times 0.0431 &=& 2.153 \text{ amu} \\
51.94 \text{ amu} \times 0.8376 &=& 43.505 \text{ amu} \\
52.94 \text{ amu} \times 0.0955 &=& 5.056 \text{ amu} \\
53.94 \text{ amu} \times 0.0238 &=& 1.283 \text{ amu} \\
\hline
\text{atomic weight} &=& 51.997 \quad = 52.00 \text{ amu}
\end{array}
$$

4.61 Group IVA: C (nonmetal); Si (metalloid); Ge (metalloid); Sn (metal); and Pb (metal). The trend in this group does agree with the general rule.

4.63 The molecular formula for chloroform is $CHCl_3$.

4.65 The SO_4^{2-} ion formula means that there is one sulfur atom in the ion and four oxygen atoms in the ion, and an excess of two electrons.

■ Solutions to Practice in Problem Analysis

4.1 Write the atomic symbols for aluminum (Al) and sulfur (S) with subscripts indicating the number of aluminum atoms (2) and sulfur atoms (3). Write the symbols in the order given in the problem.

4.2 For each of the statements given, try to determine the reference to the information in the chapter. The soft elements that react with water to form hydrogen gas are the Group IA elements (which include lithium, sodium, and potassium). The statement that it combines chemically with chlorine to form a compound with relatively high melting point that dissolves in water to give a solution that conducts electricity implies that the compound formed is an ionic compound. Thus, the unknown element must be a metal. From these statements, we can conclude that the element is a Group IA metal. The exact identity cannot be determined.

■ Answers to the Practice Exam

1. c   2. b   3. c   4. a   5. b   6. e   7. c   8. d   9. a   10. b

11. e   12. c   13. d   14. b   15. a

5. CHEMICAL FORMULAS AND NAMES

Solutions to Exercises

5.1 Barium is a main group metal, so we expect a cation. Because barium is in Group IIA, the group number is 2 and the expected cation charge is +2.

5.2 a. Li^+ and F^- form LiF   b. Li^+ and S^{2-} form Li_2S c. Li^+ and N^{3-} form Li_3N

5.3 a. K_2S, potassium sulfide   b. Li_3N, lithium nitride   c. Al_2O_3, aluminum oxide

5.4 Barium sulfide contains the barium ion, Ba^{2+}, and the sulfide ion, S^{2-}. The formula is BaS. Lithium chloride contains the lithium ion, Li^+, and the chloride ion, Cl^-. The formula is LiCl.

5.5 SnO contains the tin ion, Sn^{2+}, and the oxide ion O^{2-}. According to the Stock method, the name of the compound is tin(II) oxide. Its classical name is stannous oxide. SnO_2 contains the tin ion, Sn^{4+}, and the oxide ion, O^{2-}. According to the Stock method, the name of the compound is tin(IV) oxide. Its classical name is stannic oxide.

5.6 a. K_2SO_3, potassium sulfite   b. $Ba(ClO)_2$, barium hypochlorite

c. NH_4CN, ammonium cyanide

5.7 Barium phosphate contains the barium ion, Ba^{2+}, and the phosphate ion, PO_4^{3-}. The formula of the compound is $Ba_3(PO_4)_2$.

5.8 a. P_2O_3, diphosphorus trioxide   b. ClF_3, chlorine trifluoride

c. CCl_4, carbon tetrachloride

5.9 The name for HBr(aq) is hydrobromic acid.

5.10 The oxyacid corresponding to the sulfite ion, SO_3^{2-}, is sulfurous acid. The formula is H_2SO_3.

■ Answers to Questions to Test Your Reading

5.1 A binary ionic compound is a compound composed of ions from only two elements. In most cases, these compounds consist of a metal, which is the cation (positive ion), and a nonmetal, which is the anion (negative ion). An example is sodium chloride, NaCl.

5.2 When naming a binary ionic compound, the cation is always named first.

5.3 Cations that form a single charge are named after the metallic element from which they are derived. Thus, the names of the ions are: sodium ion, potassium ion, magnesium ion, and calcium ion.

5.4 Anions are named by retaining the nonmetallic elements root name and adding the suffix -ide. Therefore, the names of the ions are fluoride, chloride, bromide, and iodide.

5.5 According to the Stock method, the names of the Cu^+ and Cu^{2+} ions are copper(I) and copper(II), respectively. Their classical names are cuprous ion and cupric ion, respectively.

5.6 According to the Stock method, the names of the Pb^{2+} and Pb^{4+} ions are lead(II) and lead(IV), respectively. Their common names are plumbous and plumbic, respectively.

5.7 A polyatomic ion is a group of chemically bonded atoms that function as a single unit. An oxyanion is a negatively charged polyatomic ion that contains an atom of some element in addition to one or more oxygen atoms. A polyatomic ion does not have to be an oxyanion. An example of a polyatomic ion that is not an oxyanion is the ammonium ion, NH_4^+. an example of a polyatomic ion that is an oxyanion is the carbonate ion, CO_3^{2-}.

5.8 A binary molecular compound is a molecular compound composed of only two elements. It differs from a binary ionic compound in that it usually contain two nonmetals, whereas a binary ionic compound contains a metal and a nonmetal.

5.9 An acid is a compound that produces hydrogen ion (H^+) when dissolved in water. A binary acid is an acid solution that forms when you dissolve a binary molecular compound of hydrogen and another nonmetallic element in water. An example of a binary acid is hydrochloric acid, HCl(aq).

5.10 An oxyacid is a molecular substance containing hydrogen, oxygen, and another element, that when added to water yields hydrogen (H^+) ion and the corresponding oxyanion. An example is nitric acid, HNO_3.

Solutions to Practice Problems

5.11 a. +1 b. −1 c. −3 d. +2 e. +2

5.13 Two Na^+ ions and one CO_3^{2-} ion.

5.15 Al_2O_3

5.17 a. MgO b. $CaCl_2$ c. Na_2O
   d. AlF_3 e. CsBr f. Zn_3N_2

5.19 a. Li^+ and O^{2-} b. Ba^{2+} and O^{2-} c. Zn^{2+} and Cl^-
   d. K^+ and P^{3-} e. Na^+ and Br^- f. Ca^{2+} and I^-

5.21 potassium iodide

5.23 a. lithium bromide b. calcium bromide c. rubidium oxide
   d. barium nitride e. radium sulfide f. aluminum fluoride

5.25 a. BaO b. $SrCl_2$ c. Na_2S
   d. NaBr e. Al_2S_3 f. K_3N

5.27 a. cobalt(III) oxide, cobaltic oxide b. cobalt(II) oxide, cobaltous oxide
   c. lead(IV) oxide, plumbic oxide d. lead(II) oxide, plumbous oxide
   e. copper(II) oxide, cupric oxide f. copper(I) oxide, cuprous oxide

5.29 a. $SnCl_4$ is tin(IV) chloride, or stannic chloride.
   b. correct
   c. $CuBr_2$ is copper(II) bromide, or cupric bromide.
   d. Fe_2O_3 is iron(III) oxide, or ferric oxide.
   e. correct
   f. $CaCl_2$ is calcium chloride.

5.31 $CaCO_3$, calcium carbonate

CHEMICAL FORMULAS AND NAMES

5.33 a. Ca(HCO$_3$)$_2$, calcium hydrogen carbonate   b. NH$_4$Cl, ammonium chloride
c. Mg(ClO)$_2$, magnesium hypochlorite   d. AgCN, silver cyanide
e. NH$_4$NO$_3$, ammonium nitrate   f. Al(ClO$_4$)$_3$, aluminum perchlorate
g. Ba$_3$(PO$_4$)$_2$, barium phosphate   h. KNO$_2$, potassium nitrite
i. NaC$_2$H$_3$O$_2$, sodium acetate

5.35 a. Al(OH)$_3$ b. BaSO$_3$ c. BaSO$_4$
d. Li$_2$HPO$_4$ e. LiH$_2$PO$_4$ f. Li$_3$PO$_4$

5.37 nitrogen trihydride

5.39 a. N$_2$O, dinitrogen monoxide   b. CCl$_4$, carbon tetrachloride
c. SiF$_4$, silicon tetrafluoride   d. ClF$_3$, chlorine trifluoride
e. Cl$_2$O$_7$, dichlorine heptoxide

5.41 a. NBr$_3$ b. XeO$_4$ c. OF$_2$
d. Cl$_2$O$_5$ e. SF$_6$ f. PI$_3$

5.43 As a binary compound, H$_2$S is named hydrogen sulfide. As an acid, H$_2$S(aq) is named hydrosulfuric acid.

5.45 a. phosphoric acid   b. nitrous acid   c. carbonic acid
d. nitric acid   e. hypochlorous acid   f. perchloric acid

5.47 a. HNO$_2$ b. HNO$_3$ c. HI
d. H$_2$SO$_4$ e. H$_2$SO$_3$ f. HBr

■ Solutions to Additional Problems

5.49 NaHCO$_3$, sodium hydrogen carbonate

5.51 CaCO$_3$, calcium carbonate

5.53 a. FeSO$_4$ b. Fe$_2$(SO$_4$)$_3$ c. CuC$_2$H$_3$O$_2$ d. Cu(C$_2$H$_3$O$_2$)$_2$

5.55 a. NiCl$_2$ b. TiF$_4$ c. MnO$_2$ d. Cr$_2$S$_3$

5.57 a. Pb(NO$_3$)$_2$ b. Co(OH)$_2$ c. Cu$_3$(PO$_4$)$_2$
d. CuNO$_2$ e. Fe(ClO$_2$)$_2$ f. Sn(ClO$_4$)$_4$

5.59 a. tin(II) phosphate   b. iron(II) nitrate
c. chromium(II) sulfate   d. aluminum chlorate

5.61 a. BrO$_3^-$, bromate ion   b. N$_2$O$_2^{2-}$, hyponitrite ion
c. S$_2$O$_3^{2-}$, thiosulfate ion   d. AsO$_4^{3-}$, arsenate ion

■ Solutions to Practice in Problem Analysis

5.1   To determine the formula of radium chloride, first decide if it is a binary compound or whether it contains a polyatomic cation or anion. Since both radium and chlorine are elements, radium chloride is a binary compound. Next, determine whether the compound is ionic or molecular. Since radium is a main-group metal, the compound is ionic. To predict the formula of a binary ionic compound, it is necessary to determine the charge on the monoatomic ions. The charge on a main-group metal cation is equal to the group number. Radium is in Group IIA, and thus has a charge of +2. The charge on a monoatomic anion is equal to 8 minus the group number. Chlorine is in Group VIIA, and thus has a charge of -1. The formula of the ionic compound can now be determined by writing the formula of the cation first, then the anion. The subscripts in the formula are determined by the fact that the compound can have no net charge. This is equivalent to using the magnitude of the charge on radium (2) as the subscript on chlorine, and the magnitude of the charge on chlorine (1) as the subscript on radium. Subscripts equal to 1 are omitted in the formula.

5.2   To name TiCl$_4$ using the Stock method, the charge on the metal cation must be determined. This is determined by using the subscripts in the compound as the smallest cation/anion charge ratio. The subscript on the anion becomes the magnitude of the charge on the cation (4), and the subscript on the cation becomes the magnitude of the charge on the anion (1). The charge on the cation is positive, and the charge on the anion is negative. Therefore, the charges are some multiple of this ratio (+4:-1). The charge on a monoatomic anion is equal to 8 minus the group number. Chlorine is in Group VIIA, and thus has a charge of -1, the same as the charge in the ratio. Thus, titanium must have a charge of +4. In the Stock method of naming, the charge on the cation is written as a Roman numeral within parentheses after the metal name. This is followed by the name of the anion.

■ Answers to the Practice Exam

1. c   2. c   3. e   4. b   5. d   6. b   7. d   8. a   9. d   10. c
11. b 12. e 13. a 14. e 15. c 16. c 17. d 18. d 19. c 20. d

6. CHEMICAL REACTIONS AND EQUATIONS

■ Solutions to Exercises

6.1 a. The reactants are H_2, hydrogen, and O_2, oxygen. The product is H_2O, water. Both reactants are gases, while the product is a liquid. The equation means that two molecules of hydrogen gas and one molecule of oxygen gas react in the presence of a platinum catalyst to form two molecules of liquid water.

b. The reactants are NaI, sodium iodide, and Cl_2, chlorine. The products are NaCl, sodium chloride, and I_2, iodine. For the reactants, sodium iodide is aqueous (dissolved in water), and chlorine is a gas. On the product side, sodium chloride and iodine are both aqueous (dissolved in water). Two formula units of aqueous sodium iodide react with one molecule of chlorine gas to form two formula units of aqueous sodium chloride and one molecule of aqueous iodine.

c. The reactants are CO, carbon monoxide, and H_2, hydrogen. The products are CH_4, methane, and H_2O, water. Both reactants and both products are gases. One molecule of carbon monoxide gas reacts with three molecules of hydrogen gas to form one molecule of methane gas and one molecule of water gas.

d. The reactant is H_2O_2, hydrogen peroxide. It is represented as an aqueous solution. The products are H_2O, water, and O_2, oxygen. The reactant is aqueous (dissolved in water). On the product side, water is a liquid and oxygen is a gas. Two molecules of aqueous hydrogen peroxide react when heated to form two molecules of liquid water and one molecule of oxygen gas.

6.2 a. An analysis shows that the equation is not balanced.

$$NH_4NO_3(s) \xrightarrow{\Delta} N_2O(g) + H_2O(g)$$

Left side: 2 nitrogen atoms
3 oxygen atoms
4 hydrogen atoms

Right side: 2 nitrogen atoms
2 oxygen atoms
2 hydrogen atoms

Note that nitrogen and hydrogen atoms each appear in only one reactant and product. Also, the nitrogen atoms are already balanced. Thus, balance the hydrogen atoms by placing a coefficient of 2 in front of water on the right side.

$$NH_4NO_3(s) \xrightarrow{\Delta} N_2O(g) + 2\,H_2O(g)$$

A check of the oxygen atoms now shows that there are three atoms on each side, so the oxygen atoms are also balanced. Also, the coefficients are the smallest whole numbers possible. The final result is

$$NH_4NO_3(s) \xrightarrow{\Delta} N_2O(g) + 2\,H_2O(g)$$

A check of the final equation shows that it is balanced:

Left side: 2 nitrogen atoms
3 oxygen atoms
4 hydrogen atoms

Right side: 2 nitrogen atoms
3 oxygen atoms
4 hydrogen atoms

b. An analysis shows that the equation is not balanced.

$$N_2O_5(g) \longrightarrow NO_2(g) + O_2(g)$$

Left side: 2 nitrogen atoms
5 oxygen atoms

Right side: 1 nitrogen atom
4 oxygen atoms

Note that nitrogen atoms appear in only one reactant and product. Thus, start by balancing the nitrogen atoms. Place a coefficient of 2 in front of NO_2 on the right side.

$$N_2O_5(g) \longrightarrow 2\,NO_2(g) + O_2(g)$$

Next, balance oxygen atoms. A check of the oxygen atoms now shows that there are five atoms on the left side and six on the right side. Balance the oxygen atoms by placing a coefficient of ½ in front of O_2 on the right.

$$N_2O_5(g) \longrightarrow 2\,NO_2(g) + \tfrac{1}{2}\,O_2(g)$$

(continued)

The equation is now balanced, but you need to multiply all of the coefficients by 2 to clear the fraction. The final result is

$$2\,N_2O_5(g) \longrightarrow 4\,NO_2(g) + O_2(g)$$

A check of the final equation shows that it is balanced:

Left side: 4 nitrogen atoms
           10 oxygen atoms

Right side: 4 nitrogen atoms
            10 oxygen atoms

c. An analysis shows that the equation is not balanced.

$$Ca(s) + H_2O(l) \longrightarrow Ca(OH)_2(aq) + H_2(g)$$

Left side: 1 calcium atom
           2 hydrogen atoms
           1 oxygen atom

Right side: 1 calcium atom
            4 hydrogen atoms
            2 oxygen atoms

Note that calcium and oxygen atoms each appear in only one reactant and product. Also, the calcium atoms are already balanced. Thus, start by balancing the oxygen atoms. Place a coefficient of 2 in front of H_2O on the left side.

$$Ca(s) + 2\,H_2O(l) \longrightarrow Ca(OH)_2(aq) + H_2(g)$$

A check of the hydrogen atoms now shows that there are four atoms on each side, so the hydrogen atoms are also balanced. Also, the coefficients are the smallest whole numbers possible. The final result is

$$Ca(s) + 2\,H_2O(l) \longrightarrow Ca(OH)_2(aq) + H_2(g)$$

A check of the final equation shows that it is balanced:

Left side: 1 calcium atom
           4 hydrogen atoms
           2 oxygen atoms

Right side: 1 calcium atom
            4 hydrogen atoms
            2 oxygen atoms

d. An analysis shows that the equation is not balanced.

$$As_2S_3(s) + O_2(g) \xrightarrow{\Delta} As_2O_3(s) + SO_2(g)$$

Left side: 2 arsenic atoms
3 sulfur atoms
2 oxygen atoms

Right side: 2 arsenic atoms
1 sulfur atom
5 oxygen atoms

Note that arsenic and sulfur atoms each appear in only one reactant and product. Also, the arsenic atoms are already balanced. Thus, balance the sulfur atoms by placing a coefficient of 3 in front of SO_2 on the right side.

$$As_2S_3(s) + O_2(g) \xrightarrow{\Delta} As_2O_3(s) + 3\,SO_2(g)$$

Next, balance oxygen atoms. A check of the oxygen atoms now shows that there are two atoms on the left side and nine on the right side. Balance the oxygen atoms by placing a coefficient of 9/2 in front of O_2 on the left.

$$As_2S_3(s) + 9/2\,O_2(g) \xrightarrow{\Delta} As_2O_3(s) + 3\,SO_2(g)$$

The equation is now balanced, but you need to multiply all of the coefficients by 2 to clear the fraction. The final result is

$$2\,As_2S_3(s) + 9\,O_2(g) \xrightarrow{\Delta} 2\,As_2O_3(s) + 6\,SO_2(g)$$

A check of the final equation shows that it is balanced:

Left side: 4 arsenic atoms
6 sulfur atoms
18 oxygen atoms

Right side: 4 arsenic atoms
6 sulfur atoms
18 oxygen atoms

6.3 Begin by writing an unbalanced equation that agrees with the description, and analyze the number of atoms of each element.

$$Fe + O_2 \longrightarrow Fe_2O_3 \quad \text{(unbalanced)}$$

Left side: 1 iron atom
2 oxygen atoms

Right side: 2 iron atoms
3 oxygen atoms

The analysis shows that the equation is not balanced. Note that both iron atoms and oxygen atoms occur in only one reactant, as well as in the only product. Begin by balancing oxygen atoms. There are two oxygen atoms on the left, so write a 2 for the coefficient of Fe_2O_3. There are three oxygen atoms on the right, so write a 3 for the coefficient of O_2. This gives

$$Fe + 3\,O_2 \longrightarrow 2\,Fe_2O_3$$

(continued)

Next, balance the iron atoms. There are four iron atoms on the right side. To balance, put a 4 in front of Fe on the left.

$$4\ Fe + 3\ O_2 \longrightarrow 2\ Fe_2O_3$$

Check that the equation is balanced.

Left side: 4 iron atoms   Right side: 4 iron atoms
            6 oxygen atoms   6 oxygen atoms

The equation is balanced. Note also that the coefficients are the smallest whole numbers possible. Finally, add the physical states of the substances.

$$4\ Fe(s) + 3\ O_2(g) \longrightarrow 2\ Fe_2O_3(s)$$

6.4 a. This is a decomposition reaction since a compound has decomposed into an element and a compound.

b. This is a combination reaction since two elements combine to form a third substance.

c. Since this reaction is not represented by either A + B → AB or AB → A + B, it is neither a combination reaction nor a decomposition reaction.

6.5 The equation: $2\ P(s) + 5\ Cl_2(g) \longrightarrow 2\ PCl_5(s)$ does not represent a single replacement reaction because it cannot be represented by A + BC → AC + B. It is a combination reaction because it can be represented by A + B → AB.

6.6 Magnesium is above silver in the activity series. Therefore, magnesium metal will replace silver in silver nitrate. The balanced chemical equation is

$$Mg(s) + 2\ AgNO_3(aq) \longrightarrow Mg(NO_3)_2(aq) + 2\ Ag(s)$$

6.7 Write the equation for the assumed double-displacement reaction:

$$(NH_4)_2CO_3(aq) + CaBr_2(aq) \longrightarrow CaCO_3(s) + NH_4Br(aq)$$

Table 6.3 tells you that calcium carbonate ($CaCO_3$) is insoluble and that NH_4Br is soluble. Thus, $CaCO_3$ will precipitate. The balanced equation for the chemical reaction is

$$(NH_4)_2CO_3(aq) + CaBr_2(aq) \longrightarrow CaCO_3(s) + 2\ NH_4Br(aq)$$

6.8 Begin by writing the reactants. Exchange the anions and write the correct formulas for the products. This gives the following balanced chemical equation.

$$CaCO_3 + 2\ HNO_3 \longrightarrow Ca(NO_3)_2 + H_2CO_3$$

However, according to Table 6.4, carbonic acid is unstable and decomposes to give carbon dioxide and water. This gives

$$CaCO_3 + 2\ HNO_3 \longrightarrow Ca(NO_3)_2 + CO_2 + H_2O$$

To finish, add labels for the states of the substances. Referring to Table 6.3, calcium carbonate is insoluble, nitric acid is an aqueous solution, and calcium nitrate is soluble. Finally, carbon dioxide is a gas and water is a liquid. The result is

$$CaCO_3(s) + 2\ HNO_3(aq) \longrightarrow Ca(NO_3)_2(aq) + CO_2(g) + H_2O(l)$$

6.9 Start by writing the formulas of the reactants, sulfuric acid (H_2SO_4) and calcium hydroxide ($Ca(OH)_2$). Exchange the anions and obtain the products, calcium sulfate ($CaSO_4$) and water (H_2O). The balanced chemical equation is

$$H_2SO_4(aq) + Ca(OH)_2(aq) \longrightarrow CaSO_4(s) + 2\ H_2O(l)$$

The compound that forms is calcium sulfate.

6.10 a. The balanced chemical equation is

$$C_3H_8(g) + 5\ O_2(g) \longrightarrow 3\ CO_2(g) + 4\ H_2O(g)$$

This is a combustion reaction because of the reaction with oxygen (O_2) and the formation of CO_2 and H_2O as products.

b. The balanced equation is

$$Mg(s) + 2\ HBr(aq) \longrightarrow MgBr_2(aq) + H_2(g)$$

This is a single replacement reaction because the element magnesium (Mg) is reacting with a compound, HBr, and is replacing the element hydrogen (H) in the compound.

c. The balanced equation is

$$CS_2(l) + 3\ O_2(g) \longrightarrow CO_2(g) + 2\ SO_2(g)$$

This may be a combustion reaction because of the reaction with oxygen (O_2).

Answers to Questions to Test Your Reading

6.1 A chemical reaction is a change in which one or more kinds of matter are transformed into one or more new kinds of matter. A reactant is a chemical substance involved at the start of a chemical reaction. A product is a chemical substance that result from a chemical reaction.

6.2 A chemical equation is a symbolic way of expressing a chemical reaction. A coefficient is a number in front of a formula in the chemical equation. For a molecular reaction, the coefficients (which should be whole numbers) tell you how many molecules of each reactant are involved and how many molecules of each product are formed. An example of a balanced equation is

$$CH_4 + 2\,O_2 \longrightarrow CO_2 + 2\,H_2O$$

In this example, the equation means that one molecule of methane and two molecules of oxygen react to form one molecule of carbon dioxide and two molecules of water.

6.3 A combination reaction is a reaction in which two substances chemically combine to form a third. An example is the balanced equation describing the overall formation of rust.

$$4\,Fe(s) + 3\,O_2(g) \longrightarrow 2\,Fe_2O_3(s)$$

This is a combination reaction because iron chemically combines with oxygen to form a third substance, iron (III) oxide.

6.4 A decomposition reaction is a reaction in which a single compound breaks up into two or more other substances. The products of a decomposition reaction may be two or more elements, but it is not a requirement. For example, the decomposition of mercury(II) oxide forms two elements.

$$2\,HgO(s) \xrightarrow{\Delta} 2\,Hg(l) + O_2(g)$$

However, the decomposition of limestone produces two compounds.

$$2\,CaCO_3(s) \xrightarrow{\Delta} CaO(s) + CO_2(g)$$

6.5 A single replacement reaction is a reaction in which an element reacts with a compound and replaces another element in the compound. An example is the reaction of metallic zinc with hydrochloric acid.

$$Zn(s) + 2\,HCl(aq) \longrightarrow ZnCl_2(aq) + H_2(g)$$

Zinc replaces hydrogen in HCl to give $ZnCl_2$. Hydrogen is released and appears as the element H_2.

6.6 The activity series (Table 6.2) is a list of metallic elements (plus hydrogen) ordered by their relative activities in single-replacement reactions to form a given ion in aqueous solution. An element A that is above another element B in the activity series will replace that element B from its compound BC.

6.7 A double replacement reaction is a type of reaction in which two compounds exchange parts to form two new compounds. Reactions of this type have also been called metathesis reactions. There is no difference.

6.8 The three different ways that double replacement reactions can be driven in the direction of the arrow are the formation of a precipitate, the formation of a gaseous product, and the formation of an especially stable molecular substance that then remains in the solution, such as water. An example of the formation of a precipitate is the reaction of potassium iodide and lead(II) nitrate solutions to form solid lead(II) iodide.

$$2\,KI(aq) + Pb(NO_3)_2(aq) \longrightarrow PbI_2(s) + 2\,KNO_3(aq)$$

An example of the formation of a gaseous product is the reaction of acetic acid with baking soda to form carbon dioxide gas.

$$HC_2H_3O_2(aq) + NaHCO_3(aq) \longrightarrow NaC_2H_3O_2(aq) + CO_2(g) + H_2O(l)$$

An example of a reaction in which water forms is the reaction of hydrochloric acid with sodium hydroxide solution.

$$HCl(aq) + NaOH(aq) \longrightarrow H_2O(l) + NaCl(aq)$$

6.9 An acid is a compound that produces hydrogen ions (H^+) when it is dissolved in water. A base is a compound that produces hydroxide ions (OH^-) when it is dissolved in water. A neutralization reaction is the reaction that occurs between an acid and a base with the formation of an ionic compound and usually water. An example is

$$HCl(aq) + NaOH(aq) \longrightarrow H_2O(l) + NaCl(aq)$$

CHAPTER 6

6.10 A combustion reaction is a reaction of a substance with either pure oxygen or oxygen in the air with the with the rapid release of heat and the appearance of a flame. Not all combustion reactions can be classified as combination reactions, but some can be. An example of a combustion reaction that is also a combination reaction is the combustion of coal, or carbon.

$$C(s) + O_2(g) \longrightarrow CO_2(g)$$

An example of a combustion reaction that is not a combination reaction is the combustion of methane.

$$CH_4(g) + 2\,O_2(g) \longrightarrow CO_2(g) + 2\,H_2O(g)$$

■ Solutions to Practice Problems

6.11 a. Two atoms of solid potassium and one molecule of liquid bromine react to form two formula units of solid potassium bromide.

b. One molecule of solid phosphorus and six molecules of chlorine gas react to form four molecules of liquid phosphorus trichloride.

6.13 a. One formula unit of solid barium carbonate reacts when heated to produce one formula unit of solid barium oxide and one molecule of carbon dioxide gas.

b. Two molecules of aqueous hydrogen peroxide react in the presence of a potassium iodide catalyst to produce two molecules of liquid water and one molecule of oxygen gas.

6.15 a. $2\,CH_3OH + 3\,O_2 \longrightarrow 2\,CO_2 + 4\,H_2O$

b. $2\,Mg + SiO_2 \longrightarrow 2\,MgO + Si$

c. $Cl_2O_7 + H_2O \longrightarrow 2\,HClO_4$

d. $TiCl_4 + 2\,H_2O \longrightarrow TiO_2 + 4\,HCl$

6.17 a. $CO + 3\,H_2 \longrightarrow CH_4 + H_2O$

b. $Ba + 2\,H_2O \longrightarrow Ba(OH)_2 + H_2$

c. $2\,H_2S + 3\,O_2 \longrightarrow 2\,H_2O + 2\,SO_2$

d. $4\,H_3PO_3 \longrightarrow 3\,H_3PO_4 + PH_3$

6.19 $N_2(g) + O_2(g) \longrightarrow 2 NO(g)$

6.21 $4 NO_2(g) + O_2(g) \longrightarrow 2 N_2O_5(g)$

6.23 a. Neither a combination nor a decomposition reaction
b. Combination reaction
c. Decomposition reaction
d. Combination reaction

6.25 a. $2 CH_4O(l) + 3 O_2(g) \longrightarrow 2 CO_2(g) + 4 H_2O(g)$
Neither a decomposition reaction nor a combination reaction

b. $2 NO_2(g) \longrightarrow 2 NO(g) + O_2(g)$
Decomposition reaction

c. $N_2(g) + 3 H_2(g) \longrightarrow 2 NH_3(g)$
Combination reaction

d. $4 Li(s) + O_2(g) \longrightarrow 2 Li_2O(g)$
Combination reaction

6.27 a. $Cd(s) + 2 AgNO_3(aq) \longrightarrow Cd(NO_3)_2(aq) + 2 Ag(s)$
Single replacement reaction

b. $C_3H_6(g) + H_2(g) \longrightarrow C_3H_8(g)$
Not a single replacement reaction

6.29 a. Since cadmium is above nickel on the activity series, the reaction will not occur as written.

b. Since aluminum is above cadmium on the activity series, the reaction will occur as written.

6.31 Exchange cations and anions so that the products can be determined. In this case, the products would be sodium nitrate and lead carbonate. According to Table 6.3, lead(II) carbonate is insoluble, and will form a precipitate. Also, sodium nitrate is soluble, according to the table. The reaction is

$$Pb(NO_3)_2(aq) + Na_2CO_3(aq) \longrightarrow PbCO_3(s) + 2 NaNO_3(aq)$$

6.33 a. $MgSO_4(aq) + 2\,NaOH(aq) \longrightarrow Mg(OH)_2(s) + Na_2SO_4(aq)$
The precipitate is $Mg(OH)_2$.

b. No precipitate will form.

c. No precipitate will form.

d. No precipitate will form.

6.35 $Na_2CO_3(aq) + 2\,HC_2H_3O_2(aq) \longrightarrow 2\,NaC_2H_3O_2(aq) + CO_2(g) + H_2O(l)$
The gas that forms is CO_2.

6.37 a. No gas will form.

b. No gas will form.

c. $(NH_4)_2CO_3(aq) + 2\,HCl(aq) \longrightarrow 2\,NH_4Cl(aq) + CO_2(g) + H_2O(l)$
The gas that forms is CO_2.

d. No gas will form.

6.39 a. base   b. acid   c. base   d. acid   e. neither
f. neither   g. neither   h. base   i. neither

6.41 a. $HNO_3(aq) + NaOH(aq) \longrightarrow NaNO_3(aq) + H_2O(l)$

b. $Ba(OH)_2(aq) + 2\,HCl(aq) \longrightarrow BaCl_2(aq) + 2\,H_2O(l)$

c. $LiOH(aq) + HBr(aq) \longrightarrow LiBr(aq) + H_2O(l)$

d. $HC_2H_3O_2(aq) + NaOH(aq) \longrightarrow NaC_2H_3O_2(aq) + H_2O(l)$

6.43 The reaction of octane (C_8H_{18}) with oxygen is a combustion reaction. The balanced equation to form only carbon dioxide is

$$2\,C_8H_{18}(l) + 25\,O_2(g) \longrightarrow 16\,CO_2(g) + 18\,H_2O(g)$$

The balanced equation to form only carbon monoxide is

$$2\,C_8H_{18}(l) + 17\,O_2(g) \longrightarrow 16\,CO_2(g) + 18\,H_2O(g)$$

Solutions to Additional Problems

6.45 $NH_4Cl(s) \xrightarrow{\Delta} NH_3(g) + HCl(g)$

6.47 $6\ CO_2 + 6\ H_2O \longrightarrow C_6H_{12}O_6 + 6\ O_2$

6.49 $2\ NaC_{18}H_{36}O_2(aq) + CaCl_2(aq) \longrightarrow Ca(C_{18}H_{36}O_2)_2(s) + 2\ NaCl(aq)$

6.51 $6\ Li(s) + N_2(g) \longrightarrow 2\ Li_3N(s)$

6.53 a. $HNO_3(aq) + NaOH(aq) \longrightarrow NaNO_3(aq) + H_2O(l)$
The reaction is driven by the formation of water.

b. $HNO_3(aq) + NaCN(aq) \longrightarrow NaNO_3(aq) + HCN(g)$
The reaction is driven by the formation of a gas.

c. $Pb(NO_3)_2(aq) + 2\ NaCl(aq) \longrightarrow 2\ NaNO_3(aq) + PbCl_2(s)$
The reaction is driven by the formation of a precipitate.

6.55 a. soluble   b. soluble   c. insoluble   d. insoluble   e. insoluble
f. soluble   g. insoluble   h. soluble   i. insoluble

6.57 a. Single replacement reaction:

$2\ Na(s) + 2\ H_2O(l) \longrightarrow 2\ NaOH(aq) + H_2(g)$

b. Double replacement reaction:

$CrCl_3(aq) + 3\ NaOH(aq) \longrightarrow Cr(OH)_3(s) + 3\ NaCl(aq)$

c. Single replacement reaction:

$Zn(s) + CuSO_4(aq) \longrightarrow ZnSO_4(aq) + Cu(s)$

d. Combustion reaction:

$C_6H_{12}(l) + 9\ O_2(g) \longrightarrow 6\ CO_2(g) + 6\ H_2O(g)$

6.59 a. $BaCO_3(s) + 2\, HNO_3(aq) \longrightarrow Ba(NO_3)_2(aq) + CO_2(g) + H_2O(l)$

b. $BaCl_2(aq) + H_2SO_4(aq) \longrightarrow BaSO_4(s) + 2\, HCl(aq)$

c. $Ca(OH)_2(s) + 2\, HC_2H_3O_2(aq) \longrightarrow Ca(C_2H_3O_2)_2(aq) + 2\, H_2O(l)$

d. $2\, Al(s) + 3\, NiSO_4(aq) \longrightarrow Al_2(SO_4)_3(aq) + 3\, Ni(s)$

6.61 a. $2\, Li(s) + Cl_2(g) \longrightarrow 2\, LiCl(s)$

b. $2\, C_{10}H_{22}(s) + 31\, O_2(g) \longrightarrow 20\, CO_2(g) + 22\, H_2O(g)$

c. The combustion of decane ($C_{10}H_{22}$), in the presence of a deficiency of oxygen may form the following products: carbon (soot), carbon monoxide (CO), carbon dioxide (CO_2), and water (H_2O). The proportions of each product formed cannot be specified.

d. $2\, Au_2O_3(s) \xrightarrow{\Delta} 4\, Au(s) + 3\, O_2(g)$

6.63 $2\, PbS(s) + 3\, O_2(g) \xrightarrow{\Delta} 2\, PbO(s) + 2\, SO_2(g)$

6.65 $Fe_2O_3(s) + 3\, CO(g) \xrightarrow{\Delta} 2\, Fe(l) + 3\, CO_2(g)$

■ Solutions to Practice in Problem Analysis

6.1 To prepare barium sulfate ($BaSO_4$) in a precipitation reaction using potassium sulfate (K_2SO_4), it is necessary to use a compound that contains the barium ion (Ba^{2+}) and an anion that acts as a spectator ion with the potassium ion. Since barium sulfate is insoluble in water, it can be formed as the precipitate, then filtered and dried. The reaction can then be written by placing the reactants on the left hand side of the equation, and the products on the right hand side. An example of a compound that can be used with potassium sulfate is barium nitrate $Ba(NO_3)_2$.

6.2 To prepare sodium hydroxide (NaOH) in a precipitation reaction using calcium hydroxide ($Ca(OH)_2$), it is necessary to use a compound that contains the sodium ion (Na^+) and an anion that will form a precipitate with calcium ion. Since sodium hydroxide is soluble in water, it will be necessary to remove the calcium ions from the solution in the precipitate. The remaining solution, containing sodium ions and hydroxide ions can then be evaporated to dryness, and the sodium hydroxide recovered. The reaction can be written by placing the reactants on the left hand side of the equation, and the products on the right hand side. An example of a compound that can be used with calcium hydroxide is sodium sulfate (Na_2SO_4). The precipitate that would form is calcium sulfate, $CaSO_4$.

Answers to the Practice Exam

1. b   2. d   3. a   4. c   5. b   6. c   7. a   8. d   9. c   10. a
11. c   12. a   13. e   14. c   15. e

7. CHEMICAL COMPOSITION

■ Solutions to Exercises

Note on significant figures: When the final answer is written for the first time, it is written with one nonsignificant digit, and the least significant digit is underlined. The final answer is then rounded to the correct number of significant figures. Also, in these problems, AW stands for atomic weight.

7.1 The molecular weight of $C_2H_2F_4$ is the sum of the atomic weights of the atoms in the formula of the compound. This is

2 x AW for carbon	=	2 x 12.01 amu	=	24.02 amu
2 x AW for hydrogen	=	2 x 1.008 amu	=	2.016 amu
4 x AW for fluorine	=	4 x 19.00 amu	=	76.00 amu
	Molecular weight of $C_2H_2F_4$		=	102.03$\underline{6}$ = 102.04 amu

7.2 $Mg(OH)_2$ has the formula weight

1 x AW for magnesium	=	1 x 24.31 amu	=	24.31 amu
2 x AW for oxygen	=	2 x 16.00 amu	=	32.00 amu
2 x AW for hydrogen	=	2 x 1.008 amu	=	2.016 amu
	Molecular weight of $Mg(OH)_2$		=	58.3$\underline{2}$6 = 58.33 amu

PCl_5 has the formula weight

1 x AW for phosphorus	=	1 x 30.97 amu	=	30.97 amu
5 x AW for chlorine	=	5 x 35.45 amu	=	177.25 amu
	Molecular weight of PCl_5		=	208.22 amu

7.3 Step 1. Begin the problem by writing the factor that converts moles to molecules.

$$\frac{6.022 \times 10^{23} \text{ molecules } C_6H_{12}O_6}{1 \text{ mol } C_6H_{12}O_6}$$

Multiply the given quantity (0.225 mol glucose) by the conversion factor.

$$0.225 \text{ mol } C_6H_{12}O_6 \times \frac{6.022 \times 10^{23} C_6H_{12}O_6 \text{ molecules}}{1 \text{ mol } C_6H_{12}O_6} = 1.354 \times 10^{23}$$

$$= 1.35 \times 10^{23} C_6H_{12}O_6 \text{ molecules}$$

Step 2. The next conversions are to convert molecules to individual atoms. There are 6 carbon atoms, 12 hydrogen atoms, and 6 oxygen atoms in each glucose molecule. Multiply the given quantity (1.36×10^{23} molecules glucose) by the conversion factor for each element.

$$1.354 \times 10^{23} C_6H_{12}O_6 \text{ molecules} \times \frac{6 \text{ C atoms}}{1 C_6H_{12}O_6 \text{ molecule}} = 8.13 \times 10^{23} \text{ C atoms}$$

$$1.354 \times 10^{23} C_6H_{12}O_6 \text{ molecules} \times \frac{12 \text{ H atoms}}{1 C_6H_{12}O_6 \text{ molecule}} = 1.63 \times 10^{24} \text{ H atoms}$$

$$1.354 \times 10^{23} C_6H_{12}O_6 \text{ molecules} \times \frac{6 \text{ O atoms}}{1 C_6H_{12}O_6 \text{ molecule}} = 8.13 \times 10^{23} \text{ O atoms}$$

7.4 Iodine occurs as a diatomic molecule. The molar mass of iodine is double its atomic weight expressed in grams, or

$2 \times$ AW of I $= 2 \times 126.9$ g $= 253.8$ g

Radium is a Group IIA element. Its molar mass is its atomic weight expressed in grams, or 226.0 g.

7.5 The molar mass of sodium phosphate, Na_3PO_4, is its formula weight expressed in grams, or

3 x AW for sodium	=	3 x 22.99 amu	=	68.97 amu
1 x AW for phosphorus	=	1 x 30.97 amu	=	30.97 amu
4 x AW for oxygen	=	4 x 16.00 amu	=	64.00 amu
Formula weight of Na_3PO_4			=	163.94 amu

Thus, the molar mass of Na_3PO_4 is 163.94 g.

48 ■ CHAPTER 7

7.6 Magnesium iodide, MgI_2, is an ionic compound. Its formula weight is

1 x AW for magnesium =	1 x 24.31 amu	=	24.31 amu
2 x AW for iodine =	2 x 126.90 amu	=	253.80 amu
	Formula weight of MgI_2	=	278.11 amu

Thus, the molar mass of MgI_2 is 278.11 g. The quantity of magnesium iodide present in the sample is

$$53.8 \text{ g } MgI_2 \times \frac{1 \text{ mol } MgI_2}{278.11 \text{ g } MgI_2} = 0.193\underline{4} = 0.193 \text{ mol } MgI_2$$

7.7 Water, H_2O, is a molecule. Its molecular weight is

2 x AW for hydrogen =	2 x 1.008 amu	=	2.016 amu
1 x AW for oxygen =	1 x 16.00 amu	=	16.00 amu
	Molecular weight of H_2O	=	18.016 amu

Thus, the molar mass of water is 18.016 g. The mass of water in the drop is

$$0.00278 \text{ mol } H_2O \times \frac{18.016 \text{ g } H_2O}{1 \text{ mol } H_2O} = 0.0500\underline{8} = 0.0501 \text{ g } H_2O$$

7.8 Step 1. The molar mass of water is 18.016 g (see Exercise 7.7 for the calculation).

Step 2. Every 18.016 g of water contains 2.016 g of hydrogen and 16.00 g of oxygen.

Step 3. The mass percentages of each element are

$$\text{mass percent of hydrogen} = \frac{2.016 \text{ g}}{18.016 \text{ g}} \times 100\% = 11.1\underline{9} = 11.2\%$$

$$\text{mass percent of oxygen} = \frac{16.00 \text{ g}}{18.016 \text{ g}} \times 100\% = 88.\underline{81} = 88.8\%$$

7.9 Step 1. Calculate the molar mass of each compound. For NH_4NO_3, the formula weight is

2 × AW for nitrogen	=	2 × 14.01 amu	=	28.02 amu
4 × AW for hydrogen	=	4 × 1.008 amu	=	4.032 amu
3 × AW for oxygen	=	3 × 16.00 amu	=	48.00 amu
		Formula weight of NH_4NO_3	=	80.052 amu

Thus, the molar mass of NH_4NO_3 is 80.052 g.

For NH_2CONH_2, the formula weight is

2 × AW for nitrogen	=	2 × 14.01 amu	=	28.02 amu
4 × AW for hydrogen	=	4 × 1.008 amu	=	4.032 amu
1 × AW for carbon	=	1 × 12.01 amu	=	12.01 amu
1 × AW for oxygen	=	1 × 16.00 amu	=	16.00 amu
		Formula weight of NH_2CONH_2	=	60.062 amu

Thus, the molar mass of NH_2CONH_2 is 60.062 g.

Step 2. Decide how many grams of nitrogen are in each molar mass. In this case, both compounds contain 28.02 g of nitrogen per mole.

Step 3. Calculate the mass percentage of nitrogen in each compound.

$$\text{mass percent nitrogen in } NH_4NO_3 = \frac{28.02 \text{ g}}{80.052 \text{ g}} \times 100\% = 35.002 = 35.00\%$$

$$\text{mass percent nitrogen in } NH_2CONH_2 = \frac{28.02 \text{ g}}{60.062 \text{ g}} \times 100\% = 46.652 = 46.65\%$$

Therefore, NH_2CONH_2 contains more nitrogen per gram.

7.10 Step 1. In 100 g of methanol there is 37.5 g of carbon, 12.6 g of hydrogen, and 49.9 g of oxygen.

Step 2. Convert these quantities to moles using the molar masses.

$$\text{moles of carbon} = 37.5 \text{ g C} \times \frac{1 \text{ mol C}}{12.01 \text{ g C}} = 3.122 \text{ mol C}$$

$$\text{moles of hydrogen} = 12.6 \text{ g H} \times \frac{1 \text{ mol H}}{1.008 \text{ g H}} = 12.50 \text{ mol H}$$

$$\text{moles of oxygen} = 49.9 \text{ g O} \times \frac{1 \text{ mol O}}{16.00 \text{ g O}} = 3.119 \text{ mol O}$$

Step 3. Divide each quantity by the smallest number of moles to obtain the subscripts.

$$\text{subscript for carbon} = \frac{3.122 \text{ mol}}{3.119 \text{ mol}} = 1.00$$

$$\text{subscript for hydrogen} = \frac{12.50 \text{ mol}}{3.119 \text{ mol}} = 4.01$$

$$\text{subscript for oxygen} = \frac{3.119 \text{ mol}}{3.119 \text{ mol}} = 1.00$$

Since each subscript is a whole number, the empirical formula of methanol is CH_4O.

7.11 First, calculate the empirical formula weight of C_3H_6O.

```
3 x AW for carbon    = 3 x 12.01 amu  =  36.03 amu
6 x AW for hydrogen  = 6 x  1.008 amu =   6.048 amu
1 x AW for oxygen    = 1 x 16.00 amu  =  16.00 amu
         Empirical formula weight of C3H6O  =  58.078 amu
```

Next, calculate n.

$$n = \frac{\text{molecular weight}}{\text{empirical formula weight}} = \frac{116 \text{ amu}}{58.078 \text{ amu}} = 2.00$$

Thus, the molecular formula of ethyl acetate is $(C_3H_6O)_2$, or $C_6H_{12}O_2$.

7.12 First, find the empirical formula of the compound.

Step 1. 100 g of the compound contains 92.2 g carbon and 7.8 g of hydrogen.

Step 2. Convert to moles using the molar masses as conversion factors.

$$\text{mol C} = 92.2 \text{ g C} \times \frac{1 \text{ mol C}}{12.01 \text{ g C}} = 7.677 \text{ mol C}$$

$$\text{mol H} = 7.8 \text{ g H} \times \frac{1 \text{ mol H}}{1.008 \text{ g H}} = 7.74 \text{ mol H}$$

Step 3. Divide each quantity by the smallest number of moles to obtain the subscripts.

$$\text{subscript for carbon} = \frac{7.677 \text{ mol}}{7.677 \text{ mol}} = 1.00$$

$$\text{subscript for hydrogen} = \frac{7.74 \text{ mol}}{7.677 \text{ mol}} = 1.01$$

Since the subscripts are both whole numbers, the empirical formula is CH, with an empirical formula weight of

```
1 x AW for carbon     =  1 x   12.01 amu   =   12.01 amu
1 x AW for hydrogen   =  1 x    1.008 amu  =    1.008 amu
            Empirical formula weight of CH =   13.018 amu
```

Since the molecular weight is 78.1 amu, you can calculate n.

$$n = \frac{\text{molecular weight}}{\text{empirical formula weight}} = \frac{78.1 \text{ amu}}{13.018 \text{ amu}} = 6.00$$

Thus, the molecular formula of the compound is $(CH)_6$, or C_6H_6.

■ Answers to Questions to Test Your Reading

7.1 Formula weight is the sum of the atomic weights of all of the atoms in a formula unit of the substance. It can be used with either molecular compounds or ionic compounds. The molecular weight of a compound is the sum of the atomic weights of the atoms in the compound. This term normally applies to molecular compounds.

7.2 Molecular compounds can be expressed as either a molecular weight or a formula weight, so molecular compounds can have both. Ionic compounds must be expressed in formula weights only.

52 ■ CHAPTER 7

7.3 Avogadro's number is the number of atoms in exactly 12 g of carbon-12. This number is 6.022137×10^{23}. Therefore, Avogadro's number of chlorine atoms is 6.022137×10^{23} chlorine atoms, and Avogadro's number of diatomic chlorine molecules is 6.022137×10^{23} chlorine molecules.

7.4 A mole (abbreviated mol) is the quantity of a substance that contains Avogadros number (6.022×10^{23}) of atoms, molecules, or formula units of that substance (as expressed by its formula). The molar mass of a substance is the mass of one mole of the substance, in grams. The molar mass of lithium is 6.941 g. The molar mass of oxygen, O_2, is 2×16.00 g = 32.00 g.

7.5 One mole of diatomic oxygen (O_2) contains Avogadro's number, 6.022×10^{23}, oxygen molecules. Since each oxygen molecule contains two oxygen atoms, there are $2 \times 6.022 \times 10^{23}$, or 1.204×10^{24} oxygen atoms in a mole of diatomic oxygen molecules.

7.6 One mole of ozone (O_3) contains 6.022×10^{23} ozone molecules. Since each ozone molecule contains three oxygen atoms, there are $3 \times 6.022 \times 10^{23}$, or 1.807×10^{24} oxygen atoms in a mole of ozone.

7.7 Use the relation (empirical formula)$_n$ = molecular formula. To determine n, use

$$n = \frac{\text{molecular weight}}{\text{empirical formula weight}}$$

The empirical formula of hydrogen peroxide is HO, with empirical formula weight 1.008 amu + 16.00 amu = 17.10 amu. Thus,

$$n = \frac{34 \text{ amu}}{17.01 \text{ amu}} = 2.00$$

Thus, the molecular formula of hydrogen peroxide is (HO)$_2$, or H_2O_2.

7.8 The empirical formula of $C_6H_{12}O_6$ is obtained by dividing each subscript by the greatest common factor of the subscripts. In this case, each subscript can be divided by six. This gives the empirical formula CH_2O.

Solutions to Practice Problems

Note on significant figures: When the final answer is written for the first time, it is written with one nonsignificant digit, and the least significant digit is underlined. The final answer is then rounded to the correct number of significant figures. Also, in these problems, AW stands for atomic weight.

7.9 a. 2 x AW for fluorine = 2 x 19.00 amu = 38.00 amu

b. 1 x AW for phosphorus = 1 x 30.97 amu = 30.97 amu
    5 x AW for fluorine   = 5 x 19.00 amu = 95.00 amu
    The molecular weight of PF_5 = 125.97 amu

c. 1 x AW for sulfur = 1 x 32.07 amu = 32.07 amu
    3 x AW for oxygen = 3 x 16.00 amu = 48.00 amu
    The molecular weight of SO_3 = 80.07 amu

d. 4 x AW for hydrogen = 4 x 1.008 amu = 4.032 amu
    2 x AW for carbon   = 2 x 12.01 amu = 24.02 amu
    2 x AW for oxygen   = 2 x 16.00 amu = 32.00 amu
    The molecular weight of $HC_2H_3O_2$ = 60.05$\underline{2}$ = 60.05 amu

e. 6 x AW for carbon   = 6 x 12.01 amu = 72.06 amu
    6 x AW for hydrogen = 6 x 1.008 amu = 6.048 amu
    6 x AW for oxygen   = 6 x 16.00 amu = 96.00 amu
    The molecular weight of $C_6H_6O_6$ = 174.1$\underline{0}$8 = 174.11 amu

f. 12 x AW for carbon   = 12 x 12.01 amu = 144.12 amu
    22 x AW for hydrogen = 22 x 1.008 amu = 22.176 amu
    11 x AW for oxygen   = 11 x 16.00 amu = 176.00 amu
    The molecular weight of $C_{12}H_{22}O_{11}$ = 342.29$\underline{6}$ = 342.30 amu

7.11 For tartaric acid, $C_4H_6O_2$, the molecular weight is

    4 x AW for carbon   = 4 x 12.01 amu = 48.04 amu
    6 x AW for hydrogen = 6 x 1.008 amu = 6.048 amu
    2 x AW for oxygen   = 2 x 16.00 amu = 32.00 amu
                                          86.0$\underline{8}$8 = 86.09 amu

(continued)

54 ■ CHAPTER 7

For glucose, $C_6H_{12}O_6$, the molecular weight is

6 x AW for carbon	=	6 x 12.01 amu =	72.06 amu
12 x AW for hydrogen	=	12 x 1.008 amu =	12.096 amu
6 x AW for oxygen	=	6 x 16.00 amu =	96.00 amu
			180.156 = 180.16 amu

7.13 a.

1 x AW for zinc	=	1 x 65.38 amu	=	65.38 amu
2 x AW for iodine	=	2 x 126.90 amu	=	253.80 amu
	Formula weight of ZnI_2		=	319.18 amu

b.

1 x AW for aluminum	=	1 x 26.98 amu	=	26.98 amu
3 x AW for bromine	=	3 x 79.90 amu	=	239.70 amu
	Formula weight of $AlBr_3$		=	266.68 amu

c.

2 x AW for carbon	=	2 x 12.01 amu	=	24.02 amu
6 x AW for hydrogen	=	6 x 1.008 amu	=	6.048 amu
	Formula weight of C_2H_6		=	30.068 = 30.07 amu

d.

2 x AW for nitrogen	=	2 x 14.01 amu	=	28.02 amu
4 x AW for hydrogen	=	4 x 1.008 amu	=	4.032 amu
3 x AW for oxygen	=	3 x 16.00 amu	=	48.00 amu
	Formula weight of NH_4NO_3		=	80.052 = 80.05 amu

e.

2 x AW for iron	=	2 x 55.85 amu	=	111.70 amu
3 x AW for oxygen	=	3 x 16.00 amu	=	48.00 amu
	Formula weight of Fe_2O_3		=	159.70 amu

f.

2 x AW for sodium	=	2 x 22.99 amu	=	45.98 amu
1 x AW for chromium	=	1 x 52.00 amu	=	52.00 amu
4 x AW for oxygen	=	4 x 16.00 amu	=	64.00 amu
	Formula weight of Na_2CrO_4		=	161.98 amu

7.15 The answer is $MnCl_2$ (d). The formula weights of the compounds are

a. 117.16 amu   b. 95.21 amu   c. 259.90 amu
d. 125.84 amu   e. 97.99 amu   f. 189.70 amu

CHEMICAL COMPOSITION 55

7.17 First, convert moles to molecules using Avogadro's number.

$$2.33 \text{ mol } C_2H_5SH \times \frac{6.022 \times 10^{23} \text{ } C_2H_5SH \text{ molecules}}{1 \text{ mol } C_2H_5SH} = 1.403 \times 10^{24} \text{ molecules}$$

Next, note that one molecule of C_2H_5SH contains two atoms of carbon, six atoms of hydrogen, and one atom of sulfur. Using these numbers as conversion factors gives

$$1.403 \times 10^{24} \text{ molecules} \times \frac{2 \text{ C atoms}}{1 \text{ molecule}} = 2.806 \times 10^{24} = 2.81 \times 10^{24} \text{ C atoms}$$

$$1.403 \times 10^{24} \text{ molecules} \times \frac{1 \text{ S atom}}{1 \text{ molecule}} = 1.403 \times 10^{24} = 1.40 \times 10^{24} \text{ S atoms}$$

$$1.403 \times 10^{24} \text{ molecules} \times \frac{6 \text{ H atoms}}{1 \text{ molecule}} = 8.418 \times 10^{24} = 8.42 \times 10^{24} \text{ H atoms}$$

7.19 First, convert moles of magnesium nitride, Mg_3N_2, to formula units using Avogadro's number.

$$0.234 \text{ mol } Mg_3N_2 \times \frac{6.022 \times 10^{23} \text{ } Mg_3N_2 \text{ formula units}}{1 \text{ mol } Mg_3N_2} = 1.409 \times 10^{23} \text{ formula units}$$

Next, note that one formula unit of Mg_3N_2 contains three magnesium ions (Mg^{2+}). Therefore,

$$1.409 \times 10^{23} \text{ formula units} \times \frac{3 \text{ } Mg^{2+} \text{ ions}}{1 \text{ formula unit}} = 4.227 \times 10^{23}$$

$$= 4.23 \times 10^{23} \text{ } Mg^{2+} \text{ ions}$$

7.21 Hydrogen is an element that exists as a diatomic molecule. Its molecular weight is

$2 \times 1.008 \text{ amu} = 2.016 \text{ amu}$

The molar mass of hydrogen, H_2, is 2.016 g.

Helium is an element. Its molar mass is its atomic weight expressed in grams, or 4.003 g.

56 ■ CHAPTER 7

7.23 Iron is an element. Its molar mass is its atomic weight expressed in grams, or 55.85 g.

Sulfur is an element that exists in S_8 molecules. Its molecular weight is

8 x 32.07 amu = 256.56 amu

The molar mass of sulfur, S_8, is 256.56 g.

7.25 a. The molecular weight of CH_3OH is

1 x AW for carbon	= 1 x	12.01 amu =	12.01 amu
4 x AW for hydrogen	= 4 x	1.008 amu =	4.032 amu
1 x AW for oxygen	= 1 x	16.00 amu =	16.00 amu
			32.042 = 32.04 amu

The molar mass of CH_3OH is 32.04 g.

b. The molecular weight of PF_5 is

1 x AW for phosphorus	= 1 x	30.97 amu =	30.97 amu
5 x AW for fluorine	= 5 x	19.00 amu =	95.00 amu
			125.97 amu

The molar mass of PF_5 is 125.97 g.

c. The molecular weight of SO_3 is

1 x AW for sulfur	= 1 x	32.07 amu =	32.07 amu
3 x AW for oxygen	= 3 x	16.00 amu =	48.00 amu
			80.07 amu

The molar mass of SO_3 is 80.07 g.

d. The molecular weight of $HC_2H_3O_2$ is

4 x AW for hydrogen	= 4 x	1.008 amu =	4.032 amu
2 x AW for carbon	= 2 x	12.01 amu =	24.02 amu
2 x AW for oxygen	= 2 x	16.00 amu =	32.00 amu
			60.052 = 60.05 amu

The molar mass of $HC_2H_3O_2$ is 60.05 g.

e. The molecular weight of $C_6H_{12}O_6$ is

6 x AW for carbon	= 6 x	12.01 amu =	72.06 amu
12 x AW for hydrogen	= 12 x	1.008 amu =	12.096 amu
6 x AW for oxygen	= 6 x	16.00 amu =	96.00 amu
			180.156 = 180.16 amu

The molar mass of $C_6H_{12}O_6$ is 180.16 g.

f. The molecular weight of $C_{12}H_{22}O_{11}$ is

    12 x AW for carbon   =   12 x 12.01 amu =   144.12 amu
    22 x AW for hydrogen =   22 x 1.008 amu =   22.176 amu
    11 x AW for oxygen   =   11 x 16.00 amu =   176.00 amu
                                                                                           342.296   = 342.30 amu

The molar mass of $C_{12}H_{22}O_{11}$ is 342.30 g.

7.27 a. The formula weight of LiH is

    1 x AW for lithium   =   1 x   6.941 amu =   6.941 amu
    1 x AW for hydrogen =   1 x   1.008 amu =   1.008 amu
                                                                                        7.949 amu

The molar mass of LiH is 7.949 g.

b. The formula weight of $Mg(ClO_4)_2$ is

    1 x AW for magnesium = 1 x   24.31 amu =   24.31 amu
    2 x AW for chlorine   = 2 x   35.45 amu =   70.90 amu
    8 x AW for oxygen   = 8 x   16.00 amu = 128.00 amu
                                                                                        223.21 amu

The molar mass of $Mg(ClO_4)_2$ is 223.21 g.

c. The formula weight of $VOCl_3$ is

    1 x AW for vanadium = 1 x   50.94 amu =   50.94 amu
    1 x AW for oxygen   = 1 x   16.00 amu =   16.00 amu
    3 x AW for chlorine   = 3 x   35.45 amu = 106.35 amu
                                                                                        173.29 amu

The molar mass of $VOCl_3$ is 173.29 g.

d. The formula weight of $Ba_3(PO_4)_2$ is

    3 x AW for barium   = 3 x 137.33 amu = 411.99 amu
    2 x AW for phosphorus = 2 x   30.97 amu =   61.94 amu
    8 x AW for oxygen   = 8 x   16.00 amu = 128.00 amu
                                                                                       601.93 amu

The molar mass of $Ba_3(PO_4)_2$ is 601.93 g.

58 ■ CHAPTER 7

e. The formula weight of C_3H_7Br is

3 x AW for carbon	=	3 x	12.01 amu	=	36.03 amu
7 x AW for hydrogen	=	7 x	1.008 amu	=	7.056 amu
1 x AW for bromine	=	1 x	79.90 amu	=	79.90 amu
					122.986 = 122.99 amu

The molar mass of C_3H_7Br is 122.99 g.

f. The formula weight of K_2SO_3 is

2 x AW for potassium	=	2 x	39.10 amu	=	78.20 amu
1 x AW for sulfur	=	1 x	32.07 amu	=	32.07 amu
3 x AW for oxygen	=	3 x	16.00 amu	=	48.00 amu
					158.27 amu

The molar mass of K_2SO_3 is 158.27 g.

7.29 a. The molar mass of hydrogen, H_2, is 2.016 g. Thus,

$$1.00 \text{ g } H_2 \times \frac{1 \text{ mol } H_2}{2.016 \text{ g } H_2} = 0.4960 = 0.496 \text{ mol } H_2$$

b. The molar mass of lithium, Li, is 6.941 g. Thus,

$$1.00 \text{ g Li} \times \frac{1 \text{ mol Li}}{6.941 \text{ g Li}} = 0.1441 = 0.144 \text{ mol Li}$$

c. The molar mass of sodium, Na, is 22.99 g. Thus,

$$1.00 \text{ g Na} \times \frac{1 \text{ mol Na}}{22.99 \text{ g Na}} = 0.04350 = 0.0435 \text{ mol Na}$$

d. The molar mass of potassium, K, is 39.10 g. Thus,

$$1.00 \text{ g K} \times \frac{1 \text{ mol K}}{39.10 \text{ g K}} = 0.02558 = 0.0256 \text{ mol K}$$

e. The molar mass of rubidium, Rb, is 85.47 g. Thus,

$$1.00 \text{ g Rb} \times \frac{1 \text{ mol Rb}}{85.47 \text{ g Rb}} = 0.01170 = 0.0117 \text{ mol Rb}$$

CHEMICAL COMPOSITION ■ 59

f. The molar mass of cesium, Cs, is 132.91 g. Thus,

$$1.00 \text{ g Cs} \times \frac{1 \text{ mol Cs}}{132.91 \text{ g Cs}} = 7.524 \times 10^{-3} = 7.52 \times 10^{-3} \text{ mol Cs}$$

7.31 Note: See Problem #25 for the molar mass calculations for this problem.

a. The molar mass of CH_3OH is 32.04 g. Thus,

$$5.00 \text{ g } CH_3OH \times \frac{1 \text{ mol } CH_3OH}{32.04 \text{ g } CH_3OH} = 0.1561 = 0.156 \text{ mol } CH_3OH$$

b. The molar mass of PF_5 is 125.97 g. Thus,

$$5.00 \text{ g } PF_5 \times \frac{1 \text{ mol } PF_5}{125.97 \text{ g } PF_5} = 0.03969 = 0.0397 \text{ mol } PF_5$$

c. The molar mass of SO_3 is 80.07 g. Thus,

$$5.00 \text{ g } SO_3 \times \frac{1 \text{ mol } SO_3}{80.07 \text{ g } SO_3} = 0.06244 = 0.0624 \text{ mol } SO_3$$

d. The molar mass of $HC_2H_3O_2$ is 60.03 g. Thus,

$$5.00 \text{ g } HC_2H_3O_2 \times \frac{1 \text{ mol } HC_2H_3O_2}{60.03 \text{ g } HC_2H_3O_2} = 0.08329 = 0.0833 \text{ mol } HC_2H_3O_2$$

e. The molar mass of $C_6H_{12}O_6$ is 180.16 g. Thus,

$$5.00 \text{ g } C_6H_{12}O_6 \times \frac{1 \text{ mol } C_6H_{12}O_6}{180.16 \text{ g } C_6H_{12}O_6} = 0.02775 = 0.0278 \text{ mol } C_6H_{12}O_6$$

f. The molar mass of $C_{12}H_{22}O_{11}$ is 342.30 g. Thus,

$$5.00 \text{ g } C_{12}H_{22}O_{11} \times \frac{1 \text{ mol } C_{12}H_{22}O_{11}}{342.30 \text{ g } C_{12}H_{22}O_{11}} = 0.01461 = 0.0146 \text{ mol } C_{12}H_{22}O_{11}$$

7.33 The molecular weight of hydrogen sulfide, H_2S, is

2 x AW for hydrogen	=	2 x	1.008 amu	=	2.016 amu
1 x AW for sulfur	=	1 x	32.07 amu	=	32.07 amu
					34.086 amu

Therefore, the molar mass of H_2S is 34.086 g. Thus,

$$1.11 \times 10^{-2} \text{ mol } H_2S \times \frac{34.086 \text{ g } H_2S}{1 \text{ mol } H_2S} = 0.378\underline{3} = 0.378 \text{ g } H_2S$$

7.35 The molar mass of H^+ ions is 1.008 g. Thus,

$$1.3 \times 10^{-3} \text{ mol } H^+ \times \frac{1.008 \text{ g } H^+}{1 \text{ mol } H^+} = 1.3\underline{1} \times 10^{-3} = 1.3 \times 10^{-3} \text{ g } H^+$$

7.37 First, find the molar mass of $H_2C_2O_4$. The molecular weight is

2 x AW for hydrogen	=	2 x	1.008 amu	=	2.016 amu
2 x AW for carbon	=	2 x	12.01 amu	=	24.02 amu
4 x AW for oxygen	=	4 x	16.00 amu	=	64.00 amu
					90.036 amu

Thus, the molar mass of $H_2C_2O_4$ is 90.036 g. This mass of oxalic acid contains 2.016 g of hydrogen, 24.02 g of carbon, and 64.00 g of oxygen.

Using these numbers as conversion factors gives the mass percentages.

$$\text{mass percent of hydrogen} = \frac{2.016 \text{ g}}{90.036 \text{ g}} \times 100\% = 2.239\underline{1} = 2.239\%$$

$$\text{mass percent of carbon} = \frac{24.02 \text{ g}}{90.036 \text{ g}} \times 100\% = 26.6\underline{7}8 = 26.68\%$$

$$\text{mass percent of oxygen} = \frac{64.00 \text{ g}}{90.036 \text{ g}} \times 100\% = 71.08\underline{2} = 71.08\%$$

CHEMICAL COMPOSITION ■ 61

7.39 First, find the molar mass of $C_6H_4Cl_2$. The molecular weight is

6 x AW for carbon	=	6 x	12.01 amu	=	72.06 amu
4 x AW for hydrogen	=	4 x	1.008 amu	=	4.032 amu
2 x AW for chlorine	=	2 x	35.45 amu	=	70.90 amu
					146.992 amu

Thus, the molar mass of $C_6H_4Cl_2$ is 146.992 g. This mass of para-dichlorobenzene contains 72.06 g of carbon, 4.032 g of hydrogen, and 70.90 g of chlorine.

Using these numbers as conversion factors gives the mass percentages.

$$\text{mass percent of carbon} = \frac{72.06 \text{ g}}{146.992 \text{ g}} \times 100\% = 49.023 = 49.02\%$$

$$\text{mass percent of hydrogen} = \frac{4.032 \text{ g}}{146.992 \text{ g}} \times 100\% = 2.7430 = 2.743\%$$

$$\text{mass percent of chlorine} = \frac{70.90 \text{ g}}{146.992 \text{ g}} \times 100\% = 48.234 = 48.23\%$$

7.41 First, find the mass percentage of chlorine in CsCl. The formula weight is

1 x AW for cesium	=	1 x	132.91 amu	=	132.91 amu
1 x AW for chlorine	=	1 x	35.45 amu	=	35.45 amu
					168.36 amu

Thus, the molar mass of CsCl is 168.36 g. This mass of cesium chloride contains 35.45 g of chlorine. Thus, the mass percentage of chlorine is

$$\text{mass percent of chlorine in CsCl} = \frac{35.45 \text{ g}}{168.36 \text{ g}} \times 100\% = 21.056 = 21.06\%$$

Repeat the calculation for $MgCl_2$. The formula weight is

1 x AW for magnesium	=	1 x	24.31 amu	=	24.31 amu
2 x AW for chlorine	=	2 x	35.45 amu	=	70.90 amu
					95.21 amu

(continued)

Thus, the molar mass of MgCl$_2$ is 95.21 g. This mass of magnesium chloride contains 70.90 g of chlorine. Thus, the mass percentage of chlorine is

$$\text{mass percent of chlorine in MgCl}_2 = \frac{70.90 \text{ g}}{95.21 \text{ g}} \times 100\% = 74.466 = 74.47\%$$

Since MgCl$_2$ has a higher mass percentage of chlorine, it contains more chlorine per gram of compound.

7.43 In 100 g of the compound there is 43.64 g of phosphorus and 56.36 g of oxygen. Convert these quantities to moles using the molar masses as conversion factors.

$$\text{mol phosphorus} = 43.64 \text{ g P} \times \frac{1 \text{ mol P}}{30.97 \text{ g P}} = 1.409 \text{ mol P}$$

$$\text{mol oxygen} = 56.36 \text{ g O} \times \frac{1 \text{ mol O}}{16.00 \text{ g O}} = 3.523 \text{ mol O}$$

Divide both quantities by the smallest number of moles to get the subscripts.

$$\text{subscript for phosphorus} = \frac{1.409 \text{ mol}}{1.409 \text{ mol}} = 1.000$$

$$\text{subscript for oxygen} = \frac{3.523 \text{ mol}}{1.409 \text{ mol}} = 2.500$$

The formula is now P$_{1.000}$O$_{2.500}$. You need to multiply both subscripts by two to convert of whole numbers. The empirical formula is P$_2$O$_5$.

7.45 In 100 g of the compound there is 65.4 g of carbon, 5.5 g of hydrogen, and 29.1 g of oxygen. Convert these quantities to moles using the molar masses as conversion factors.

$$\text{mol carbon} = 65.4 \text{ g C} \times \frac{1 \text{ mol C}}{12.01 \text{ g C}} = 5.45 \text{ mol C}$$

$$\text{mol hydrogen} = 5.5 \text{ g H} \times \frac{1 \text{ mol H}}{1.008 \text{ g H}} = 5.5 \text{ mol H}$$

$$\text{mol oxygen} = 29.1 \text{ g O} \times \frac{1 \text{ mol O}}{16.00 \text{ g O}} = 1.82 \text{ mol O}$$

(continued)

Divide each quantity by the smallest number of moles to get the subscripts.

$$\text{subscript for carbon} = \frac{5.45 \text{ mol}}{1.82 \text{ mol}} = 2.99$$

$$\text{subscript for hydrogen} = \frac{5.5 \text{ mol}}{1.82 \text{ mol}} = 3.0$$

$$\text{subscript for oxygen} = \frac{1.82 \text{ mol}}{1.82 \text{ mol}} = 1.00$$

Since all of the subscripts are whole numbers, the empirical formula is C_3H_3O.

7.47 The empirical formula weight of CHCl is

1 x AW for carbon	= 1 x	12.01 amu =	12.01 amu
1 x AW for hydrogen	= 1 x	1.008 amu =	1.008 amu
1 x AW for chlorine	= 1 x	35.45 amu =	35.45 amu
			48.468 amu

Find the value of n

$$n = \frac{\text{molecular weight}}{\text{empirical formula weight}} = \frac{291 \text{ amu}}{48.468 \text{ amu}} = 6.00$$

The molecular formula is $(CHCl)_6$, or $C_6H_6Cl_6$.

7.49 The empirical formula weight of P_2O_5 is

2 x AW for phosphorus =	2 x	30.97 amu =	61.94 amu
5 x AW for oxygen	= 5 x	16.00 amu =	80.00 amu
			141.94 amu

Find the value of n

$$n = \frac{\text{molecular weight}}{\text{empirical formula weight}} = \frac{284 \text{ amu}}{141.94 \text{ amu}} = 2.00$$

The molecular formula is $(P_2O_5)_2$, or P_4O_{10}.

7.51 In 100 g of the substance there is 26.7 g of carbon, 2.2 g of hydrogen, and 71.7 g of oxygen. Convert these quantities to moles using the molar masses as conversion factors.

$$\text{mol carbon} = 26.7 \text{ g C} \times \frac{1 \text{ mol C}}{12.01 \text{ g C}} = 2.22 \text{ mol C}$$

$$\text{mol hydrogen} = 2.2 \text{ g H} \times \frac{1 \text{ mol H}}{1.008 \text{ g H}} = 2.2 \text{ mol H}$$

$$\text{mol oxygen} = 71.1 \text{ g O} \times \frac{1 \text{ mol O}}{16.00 \text{ g O}} = 4.44 \text{ mol O}$$

Divide each quantity by the smallest number of moles to get the subscripts.

$$\text{subscript for carbon} = \frac{2.22 \text{ mol}}{2.2 \text{ mol}} = 1.0$$

$$\text{subscript for hydrogen} = \frac{2.2 \text{ mol}}{2.2 \text{ mol}} = 1.0$$

$$\text{subscript for oxygen} = \frac{4.44 \text{ mol}}{2.2 \text{ mol}} = 2.0$$

The subscripts are all whole numbers, so the empirical formula is CHO_2. The empirical formula weight is

$$\begin{array}{lllll}
1 \times \text{AW for carbon} & = & 1 \times 12.01 \text{ amu} & = & 12.01 \text{ amu} \\
1 \times \text{AW for hydrogen} & = & 1 \times 1.008 \text{ amu} & = & 1.008 \text{ amu} \\
2 \times \text{AW for oxygen} & = & 2 \times 16.00 \text{ amu} & = & 32.00 \text{ amu} \\
& & & & \overline{45.018 \text{ amu}}
\end{array}$$

Find the value of n

$$n = \frac{\text{molecular weight}}{\text{empirical formula weight}} = \frac{90 \text{ amu}}{45.018 \text{ amu}} = 2.0$$

Thus, the molecular formula is $(CHO_2)_2$, or $C_2H_2O_4$.

7.53 In 100 g of the substance there is 49.0 g of carbon, 2.7 g of hydrogen, and 48.2 g of chlorine. Convert these quantities to moles using the molar masses as conversion factors.

$$\text{mol carbon} = 49.0 \text{ g C} \times \frac{1 \text{ mol C}}{12.01 \text{ g C}} = 4.08 \text{ mol C}$$

$$\text{mol hydrogen} = 2.7 \text{ g H} \times \frac{1 \text{ mol H}}{1.008 \text{ g H}} = 2.7 \text{ mol H}$$

$$\text{mol chlorine} = 48.2 \text{ g Cl} \times \frac{1 \text{ mol Cl}}{35.45 \text{ g Cl}} = 1.36 \text{ mol Cl}$$

Divide each quantity by the smallest number of moles to get the subscripts.

$$\text{subscript for carbon} = \frac{4.08 \text{ mol}}{1.36 \text{ mol}} = 3.00$$

$$\text{subscript for hydrogen} = \frac{2.7 \text{ mol}}{1.36 \text{ mol}} = 2.0$$

$$\text{subscript for chlorine} = \frac{1.36 \text{ mol}}{1.36 \text{ mol}} = 1.00$$

Since all of the subscripts are whole numbers, the empirical formula is C_3H_2Cl. The empirical formula weight is

3 x AW for carbon	= 3 x 12.01 amu	=	36.03 amu
2 x AW for hydrogen	= 2 x 1.008 amu	=	2.016 amu
1 x AW for chlorine	= 1 x 35.45 amu	=	35.45 amu
			73.496 amu

Find the value of n

$$n = \frac{\text{molecular weight}}{\text{empirical formula weight}} = \frac{147 \text{ amu}}{73.496 \text{ amu}} = 2.00$$

The molecular formula is $(C_3H_2Cl)_2$, or $C_6H_4Cl_2$.

Solutions to Additional Problems

7.55 $6.3 \text{ g carbon 12} \times \dfrac{1 \text{ mol carbon 12}}{12.00 \text{ g carbon 12}} = 0.5\underline{2}5 = 0.53 \text{ mol carbon 12}$

$0.525 \text{ mol carbon 12} \times \dfrac{6.022 \times 10^{23} \text{ atoms}}{1 \text{ mol carbon 12}} = 3.1\underline{6} \times 10^{23} = 3.2 \times 10^{23} \text{ atoms}$

7.57 $2.43 \times 10^{24} \text{ atoms} \times \dfrac{1 \text{ mol carbon 12}}{6.022 \times 10^{23} \text{ atoms}} \times \dfrac{12.00 \text{ g}}{1 \text{ mol carbon 12}} = 48.\underline{4}2 = 48.4 \text{ g}$

7.59 $3.29 \times 10^{-2} \text{ mol Na}^+ \times \dfrac{22.99 \text{ g}}{1 \text{ mol Na}^+} \times \dfrac{1000 \text{ mg}}{1 \text{ g}} = 75\underline{6}.4 = 756 \text{ mg}$

7.61 The formula weight of magnesium chloride, $MgCl_2$, is

$$\begin{array}{lllll}
1 \times \text{AW for magnesium} = & 1 \times & 24.31 \text{ amu} & = & 24.31 \text{ amu} \\
2 \times \text{AW for chlorine} = & 2 \times & 35.45 \text{ amu} & = & 70.90 \text{ amu} \\
& & & & \overline{95.21 \text{ amu}}
\end{array}$$

Thus, the molar mass of $MgCl_2$ is 95.21 g.

Since there are two chloride ions per formula unit, you get

$1.11 \text{ kg MgCl}_2 \times \dfrac{1000 \text{ g}}{1 \text{ kg}} \times \dfrac{1 \text{ mol MgCl}_2}{95.21 \text{ g MgCl}_2} \times \dfrac{2 \text{ mol Cl}^-}{1 \text{ mol MgCl}_2} = 23.\underline{3}2 = 23.3 \text{ mol Cl}^-$

$23.32 \text{ mol Cl}^- \times \dfrac{6.022 \times 10^{23} \text{ Cl}^- \text{ ions}}{1 \text{ mol Cl}^- \text{ ions}} = 1.4\underline{0}4 \times 10^{25} = 1.40 \times 10^{25} \text{ Cl}^- \text{ ions}$

7.63 Since there are eight sulfur atoms per S_8 molecule, you get

$6.022 \times 10^{23} \text{ S atoms} \times \dfrac{1 \text{ S}_8 \text{ molecule}}{8 \text{ S atoms}} = 7.52\underline{7}5 \times 10^{22} = 7.528 \times 10^{22} \text{ S}_8 \text{ molecules}$

CHEMICAL COMPOSITION   67

7.65 The formula weight of calcium carbonate, $CaCO_3$, is

1 x AW for calcium	=	1 x 40.08 amu	=	40.08 amu
1 x AW for carbon	=	1 x 12.01 amu	=	12.01 amu
3 x AW for oxygen	=	3 x 16.00 amu	=	48.00 amu
				100.09 amu

Thus, the molar mass of $CaCO_3$ is 100.09 g. Since there is one calcium ion per formula unit, you get

$$3.4 \text{ g } CaCO_3 \times \frac{1 \text{ mol } CaCO_3}{100.09 \text{ g } CaCO_3} \times \frac{1 \text{ mol } Ca^{2+} \text{ ions}}{1 \text{ mol } CaCO_3} \times \frac{6.022 \times 10^{23} \text{ Ca}^{2+} \text{ ions}}{1 \text{ mol } Ca^{2+} \text{ ions}}$$

$$= 2.04 \times 10^{22} = 2.0 \times 10^{22} \text{ Ca}^{2+} \text{ ions}$$

7.67 The molecular weight of water, H_2O, is

2 x AW for hydrogen	=	2 x 1.008 amu	=	2.016 amu
1 x AW for oxygen	=	1 x 16.00 amu	=	16.00 amu
				18.016 amu

Thus, the molar mass of water is 18.016 g. Since there are two hydrogen atoms per molecule, you get

$$2.4 \text{ g } H_2O \times \frac{1 \text{ mol } H_2O}{18.016 \text{ g } H_2O} \times \frac{6.022 \times 10^{23} \text{ H}_2O \text{ molecules}}{1 \text{ mol } H_2O} \times \frac{2 \text{ H atoms}}{1 \text{ H}_2O \text{ molecule}}$$

$$= 1.60 \times 10^{23} = 1.6 \times 10^{23} \text{ H atoms}$$

7.69 In 100 g of the substance there is 58.8 b of barium, 13.8 g of sulfur, and 27.4 g of oxygen. Convert these quantities to moles using the molar masses as conversion factors.

$$\text{mol Ba} = 58.8 \text{ g Ba} \times \frac{1 \text{ mol Ba}}{137.33 \text{ g Ba}} = 0.4282 \text{ mol Ba}$$

$$\text{mol S} = 13.8 \text{ g S} \times \frac{1 \text{ mol S}}{32.07 \text{ g S}} = 0.4303 \text{ mol S}$$

$$\text{mol O} = 27.4 \text{ g O} \times \frac{1 \text{ mol O}}{16.00 \text{ g O}} = 1.713 \text{ mol O}$$

(continued)

Divide each quantity by the smallest number of moles to get the subscripts.

$$\text{subscript for Ba} = \frac{0.4282 \text{ mol}}{0.4282 \text{ mol}} = 1.00$$

$$\text{subscript for S} = \frac{0.4303 \text{ mol}}{0.4282 \text{ mol}} = 1.00$$

$$\text{subscript for O} = \frac{1.713 \text{ mol}}{0.4282 \text{ mol}} = 4.00$$

Since all of the subscripts are whole numbers, the empirical formula is $BaSO_4$.

7.71 For potassium manganate, 100 g of the compound contains 39.7 g of potassium, 27.9 g of manganese, and 32.5 g of oxygen. Convert these quantities to moles using the molar masses as conversion factors.

$$\text{mol K} = 39.7 \text{ g K} \times \frac{1 \text{ mol K}}{39.10 \text{ g K}} = 1.015 \text{ mol K}$$

$$\text{mol Mn} = 27.9 \text{ g Mn} \times \frac{1 \text{ mol Mn}}{54.94 \text{ g Mn}} = 0.5078 \text{ mol Mn}$$

$$\text{mol O} = 32.5 \text{ g O} \times \frac{1 \text{ mol O}}{16.00 \text{ g O}} = 2.031 \text{ mol O}$$

Divide each quantity by the smallest number of moles to get the subscripts.

$$\text{subscript for K} = \frac{1.015 \text{ mol}}{0.5078 \text{ mol}} = 2.00$$

$$\text{subscript for Mn} = \frac{0.5078 \text{ mol}}{0.5078 \text{ mol}} = 1.00$$

$$\text{subscript for O} = \frac{2.031 \text{ mol}}{0.5078 \text{ mol}} = 4.00$$

Since all of the subscripts are whole numbers, the empirical formula of potassium manganate is K_2MnO_4.

For potassium permanganate, 100 g of the compound contains 24.7 g of potassium, 34.8 g of manganese, and 40.5 g of oxygen. Convert these quantities to moles using the molar masses as conversion factors.

(continued)

CHEMICAL COMPOSITION ■ 69

$$\text{mol K} = 24.7 \text{ g K} \times \frac{1 \text{ mol K}}{39.10 \text{ g K}} = 0.6317 \text{ mol K}$$

$$\text{mol Mn} = 34.8 \text{ g Mn} \times \frac{1 \text{ mol Mn}}{54.94 \text{ g Mn}} = 0.6334 \text{ mol Mn}$$

$$\text{mol O} = 40.5 \text{ g O} \times \frac{1 \text{ mol O}}{16.00 \text{ g O}} = 2.531 \text{ mol O}$$

Divde each quantity by the smallest number of moles to get the subscripts.

$$\text{subscript for K} = \frac{0.6317 \text{ mol}}{0.6317 \text{ mol}} = 1.00$$

$$\text{subscript for Mn} = \frac{0.6334 \text{ mol}}{0.6317 \text{ mol}} = 1.00$$

$$\text{subscript for O} = \frac{2.531 \text{ mol}}{0.6317 \text{ mol}} = 4.01$$

Since all of the subscripts are whole numbers, the empirical formula of potassium permanganate is $KMnO_4$.

7.73   First, obtain the mass of oxygen in the compound.

mass oxygen = mass compound - mass magnesium

= 0.4145 - 0.2501 = 0.1644 g.

Thus, the sample contains 0.2501 g of magnesium and 0.1644 g of oxygen. Convert these quantities to moles using the molar masses as conversion factors.

$$\text{mol Mg} = 0.2501 \text{ g Mg} \times \frac{1 \text{ mol Mg}}{24.31 \text{ g Mg}} = 0.01029 \text{ mol Mg}$$

$$\text{mol O} = 0.1644 \text{ g O} \times \frac{1 \text{ mol O}}{16.00 \text{ g O}} = 0.01028 \text{ mol O}$$

Since both molar quantities are the same, both subscripts are equal to one, and the empirical formula is MgO.

7.75 First, obtain the mass of oxygen in the sample.

mass oxygen = mass compound - mass carbon = 2.200 - 0.600 = 1.600 g.

Thus, the sample contains 0.600 g of carbon and 1.600 g oxygen. Convert these quantities to moles using the molar masses as conversion factors.

$$\text{mol C} = 0.600 \text{ g C} \times \frac{1 \text{ mol C}}{12.01 \text{ g C}} = 0.04996 \text{ mol C}$$

$$\text{mol O} = 1.600 \text{ g O} \times \frac{1 \text{ mol O}}{16.00 \text{ g O}} = 0.1000 \text{ mol O}$$

Divide each quantity by the smaller number of moles to get the subscripts.

$$\text{subscript for C} = \frac{0.04996 \text{ mol}}{0.04996 \text{ mol}} = 1.00$$

$$\text{subscript for O} = \frac{0.1000 \text{ mol}}{0.04996 \text{ mol}} = 2.00$$

Since the subscripts are whole numbers, the formula of the compound is CO_2.

■ Solutions to Practice in Problem Analysis

7.1 To solve this problem, first you must convert the number of grams (19.4 g) of H_2SO_4 to moles using the molar mass (98.086 g/mol) as the conversion factor. Next, the number of moles can be converted to molecules of H_2SO_4 using Avogadro's number (6.022 x 10^{23} molecules/mol) as the conversion factor. Finally, to obtain the number of oxygen atoms, multiply the number of molecules by the conversion factor: 4 O atoms / 1 H_2SO_4 molecule.

7.2 First, you must determine the empirical formula of the compound. To do this, first write each mass percentage of an element as the mass in grams of the element in 100 g of the compound. Next, convert these masses to moles using the atomic weights of the elements. Then, divide each of the mole quantities by the smallest quantity. If these values are whole numbers, they are the subscripts. Otherwise, multiply all of the subscripts by some small whole number so that all of the subscripts become whole numbers. The result should be the empirical formula of the compound.

Next, to determine the actual molecular formula, calculate the empirical formula weight of the compound. This is the sum of the atomic weights of all of the atoms in the formula. Then, determine the value of n by dividing the molecular weight by the empirical formula weight. This is the factor that you multiply all of the subscripts by in order to obtain the molecular formula.

■ Answers to the Practice Exam

1. c   2. d   3. a   4. d   5. d   6. e   7. b   8. d   9. e   10. b
11. d   12. c   13. d   14. d   15. d

8. QUANTITIES IN CHEMICAL REACTIONS

■ Solutions to Exercises

Note on significant figures: When the final answer is written for the first time, it is written with one nonsignificant digit, and the least significant digit is underlined. The final answer is then rounded to the correct number of significant figures.

8.1 The calculation is

$$5.82 \text{ mol } O_2 \times \frac{4 \text{ mol Fe}}{3 \text{ mol } O_2} = 7.7\underline{6}0 = 7.76 \text{ mol Fe}$$

Thus, 7.76 moles of iron are necessary to react with 5.82 moles of oxygen.

8.2 The calculation is

$$8.4 \text{ mol } C_3H_8 \times \frac{3 \text{ mol } CO_2}{1 \text{ mol } C_3H_8} = 2\underline{5}.2 = 25 \text{ mol } CO_2$$

Thus, 25 moles of carbon dioxide will be obtained when 8.4 moles of propane are burned.

8.3 The calculation is

$$2.22 \text{ mol } H_2O \times \frac{2 \text{ mol } C_8H_{18}}{18 \text{ mol } H_2O} = 0.246\underline{7} = 0.247 \text{ mol } C_8H_{18}$$

Thus, 0.247 moles of octane are required to give 2.22 moles of water.

8.4 In this exercise, besides the mole ratio, you will also need the molar masses of iron (55.85 g) and oxygen (32.00 g). The calculation is

$$3.22 \text{ g O}_2 \times \frac{1 \text{ mol O}_2}{32.00 \text{ g O}_2} \times \frac{4 \text{ mol Fe}}{3 \text{ mol O}_2} \times \frac{55.85 \text{ g Fe}}{1 \text{ mol Fe}} = 7.4\underline{9}3 = 7.49 \text{ g Fe}$$

Therefore, 7.49 g of iron are required to react with 3.22 g of oxygen.

8.5 In this exercise, besides the mole ratio, you will also need the molar masses of **phosphorus (123.88 g) and phosphorus pentachloride (208.22 g)**. The calculation is

$$5.00 \text{ g P}_4 \times \frac{1 \text{ mol P}_4}{123.88 \text{ g P}_4} \times \frac{4 \text{ mol PCl}_5}{1 \text{ mol P}_4} \times \frac{208.22 \text{ g PCl}_5}{1 \text{ mol PCl}_5}$$

$$= 33.\underline{6}2 = 33.6 \text{ g PCl}_5$$

Thus, 33.6 g of PCl_5 can be produced from 5.00 g of P_4.

8.6 In this exercise, besides the mole ratio, you will need the molar masses of oxygen (32.00 g) and nitrogen dioxide (46.01 g). The calculation is

$$141 \text{ g NO}_2 \times \frac{1 \text{ mol NO}_2}{46.01 \text{ g NO}_2} \times \frac{1 \text{ mol O}_2}{2 \text{ mol NO}_2} \times \frac{32.00 \text{ g O}_2}{1 \text{ mol O}_2}$$

$$= 49.\underline{0}3 = 49.0 \text{ g O}_2$$

Thus, 49.0 g of oxygen are used up when 141 g of nitrogen dioxide are formed.

8.7 a. Calculate the amount of H_2 that is required to react with 27 moles of N_2.

$$27 \text{ mol N}_2 \times \frac{3 \text{ mol H}_2}{1 \text{ mol N}_2} = 81 \text{ mol H}_2$$

The available quantity of H_2 (81 mol) is exactly the right amount, so there is no limiting reactant.

b. The amount of H_2 that is required to react with 2.5 moles of N_2 is

$$2.5 \text{ mol } N_2 \times \frac{3 \text{ mol } H_2}{1 \text{ mol } N_2} = 7.5 \text{ mol } H_2$$

The available quantity of H_2 (6.2 mol) is less than the required amount, so H_2 is the limiting reactant.

c. The amount of H_2 that is required to react with 1.7 moles of N_2 is

$$1.7 \text{ mol } N_2 \times \frac{3 \text{ mol } H_2}{1 \text{ mol } N_2} = 51 \text{ mol } H_2$$

The available quantity of H_2 (9.00 mol) is more than the required amount, so N_2 is the limiting reactant.

8.8 First, determine the limiting reactant, if there is one. Find the amount of HCl that is required to react with 1.0 mole of $CaCO_3$.

$$1.0 \text{ mol } CaCO_3 \times \frac{2 \text{ mol } HCl}{1 \text{ mol } CaCO_3} = 2.0 \text{ mol } HCl$$

Since the amount of HCl available (1.0 mol) is less than the required amount, HCl is the limiting reactant. Use the available quantity of HCl to calculate the quantity of each product formed.

$$1.0 \text{ mol } HCl \times \frac{1 \text{ mol } CaCl_2}{2 \text{ mol } HCl} = 0.50 \text{ mol } CaCl_2$$

$$1.0 \text{ mol } HCl \times \frac{1 \text{ mol } CO_2}{2 \text{ mol } HCl} = 0.50 \text{ mol } CO_2$$

$$1.0 \text{ mol } HCl \times \frac{1 \text{ mol } H_2O}{2 \text{ mol } HCl} = 0.50 \text{ mol } H_2O$$

(continued)

QUANTITIES IN CHEMICAL REACTIONS ■ 75

Next, calculate the quantity of $CaCO_3$ that is used up in the reaction.

$$1.0 \text{ mol HCl} \times \frac{1 \text{ mol } CaCO_3}{2 \text{ mol HCl}} = 0.50 \text{ mol } CaCO_3$$

The quantity of $CaCO_3$ remaining after the reaction is

$$1.0 \text{ mol } CaCO_3 - 0.50 \text{ mol } CaCO_3 = 0.\underline{5}0 = 0.5 \text{ mol } CaCO_3$$

8.9 First, convert the mass of each reactant to moles, using the molar masses as conversion factors. The molar masses of $NaHCO_3$ and HCl are 84.01 g and 36.46 g, respectively. Therefore,

$$1.00 \text{ g } NaHCO_3 \times \frac{1 \text{ mol } NaHCO_3}{84.01 \text{ g } NaHCO_3} = 0.01190 \text{ mol } NaHCO_3$$

$$2.00 \text{ g HCl} \times \frac{1 \text{ mol HCl}}{36.46 \text{ g HCl}} = 0.05485 \text{ mol HCl}$$

Next, calculate the required amount of HCl necessary to react with the 0.01190 mole of $NaHCO_3$.

$$0.01190 \text{ mol } NaHCO_3 \times \frac{1 \text{ mol HCl}}{1 \text{ mol } NaHCO_3} = 0.01190 \text{ mol HCl}$$

Since the available quantity of HCl is more than the required amount, $NaHCO_3$ is the limiting reactant. Using this quantity, calculate the mass of carbon dioxide that is produced. The molar mass of CO_2 is 44.01 g. Thus,

$$0.01190 \text{ mol } NaHCO_3 \times \frac{1 \text{ mol } CO_2}{1 \text{ mol } NaHCO_3} \times \frac{44.01 \text{ g } CO_2}{1 \text{ mol } CO_2}$$

$$= 0.52\underline{3}7 = 0.524 \text{ g } CO_2$$

8.10 The reaction is

$$2\,H_2(g) + O_2(g) \longrightarrow 2\,H_2O(g)$$

First, convert the mass of hydrogen to moles using the molar mass of H_2 (2.016 g).

$$5.00\text{ g }H_2 \times \frac{1\text{ mol }H_2}{2.016\text{ g }H_2} = 2.480\text{ mol }H_2$$

Next, calculate the theoretical yield of H_2O. The molar mass of H_2O is 18.02 g. Thus,

$$2.480\text{ mol }H_2 \times \frac{2\text{ mol }H_2O}{2\text{ mol }H_2} \times \frac{18.02\text{ g }H_2O}{1\text{ mol }H_2O} = 44.69\text{ g }H_2O$$

Therefore, the percentage yield is

$$\text{percentage yield of }H_2O = \frac{0.32\text{ g}}{44.69\text{ g}} \times 100\% = 0.7\underline{1}6 = 0.72\%$$

■ Answers to Questions to Test Your Reading

8.1 Dalton's atomic theory states that atoms in molecules are neither created nor destroyed during a chemical reaction; they are merely rearranged. In a balanced equation, every atom on the left side of the equation also appears on the right side.

8.2 For the reaction

$$CO(g) + 3\,H_2(g) \longrightarrow CH_4(g) + H_2O(g)$$

In terms of molecules, one molecule of carbon monoxide (CO) reacts with three molecules of hydrogen (H_2) to form one molecule of methane (CH_4) and one molecule of water (H_2O).

In terms of moles, one mole of carbon monoxide reacts with three moles of hydrogen to form one mole of methane and two moles of water.

8.3 The chemical equation for this reaction is

$$C_2H_4(g) + 3\,O_2(g) \longrightarrow 2\,CO_2(g) + 2\,H_2O(g)$$

In terms of molecules, one molecule of ethylene (C_2H_4) reacts with three molecules of oxygen (O_2) to form two molecules of carbon dioxide (CO_2) and two molecules of water (H_2O).

In terms of moles, one mole of ethylene reacts with three moles of oxygen to form two moles of carbon dioxide and two moles of water.

8.4 The chemical equation for the reaction is

$$2\,H_2S(g) + 3\,O_2(g) \longrightarrow 2\,SO_2(g) + 2\,H_2O(g)$$

The number of moles of H_2S that will combine with 15 moles of oxygen is

$$15 \text{ mol } O_2 \times \frac{2 \text{ mol } H_2S}{3 \text{ mol } O_2} = 10.\text{ mol } H_2S$$

Thus, the statement that 5.0 mol of H_2S will combine with 15 mol of O_2 is false.

8.5 A mole ratio is a conversion factor that relates the number of moles of one of the reactants or products to the number of moles of any other reactant or product. The coefficients of the balanced chemical equation appear in the mole ratio.

8.6 Since mole ratios are conversion factors, the desired unit must be on the top and the given unit on the bottom. In this question, the given unit is moles of hydrogen and the desired unit is moles of carbon monoxide. Thus, the correct mole ratio must be

$$\frac{1 \text{ mol CO}}{3 \text{ mol } H_2}$$

8.7 In each of these reactions, the given unit (reactant) is moles of oxygen and the desired unit (product) is carbon dioxide. Thus, each mole ratio will have mol CO_2 on top and mol O_2 on the bottom.

a. $2\,CO + O_2 \longrightarrow 2\,CO_2$ The mole ratio is: $\dfrac{2 \text{ mol } CO_2}{1 \text{ mol } O_2}$

b. $CH_4 + 2 O_2 \longrightarrow CO_2 + 2 H_2O$ The mole ratio is: $\dfrac{1 \text{ mol } CO_2}{2 \text{ mol } O_2}$

c. $2 C_2H_6 + 7 O_2 \longrightarrow 4 CO_2 + 6 H_2O$ The mole ratio is: $\dfrac{4 \text{ mol } CO_2}{7 \text{ mol } O_2}$

d. $C_3H_8 + 5 O_2 \longrightarrow 3 CO_2 + 4 H_2O$ The mole ratio is: $\dfrac{3 \text{ mol } CO_2}{5 \text{ mol } O_2}$

8.8 In these reactions, moles of product will be on top of the mole ratio and moles of reactant will be on the bottom.

a. $2 SO_2 + O_2 \longrightarrow 2 SO_3$ The mole ratio is: $\dfrac{2 \text{ mol } SO_3}{2 \text{ mol } SO_2}$

b. $CS_2 + 3 Cl_2 \longrightarrow CCl_4 + S_2Cl_2$ The mole ratio is: $\dfrac{1 \text{ mol } CCl_4}{3 \text{ mol } Cl_2}$

c. $WO_3 + 3 H_2 \longrightarrow W + 3 H_2O$ The mole ratio is: $\dfrac{3 \text{ mol } H_2O}{1 \text{ mol } WO_3}$

d. $4 Li + O_2 \longrightarrow 2 Li_2O$ The mole ratio is: $\dfrac{2 \text{ mol } Li_2O}{4 \text{ mol } Li}$

8.9 The reaction in question 8.2 is

$$CO(g) + 3 H_2(g) \longrightarrow CH_4(g) + H_2O(g)$$

According to this reaction, 1 mole of carbon monoxide reacts with 3 moles of hydrogen. Coefficients in chemical reactions never refer to masses. In order to determine the mass relationships, you must use the molar masses of the substances in the calculation.

8.10 A limiting reactant is the reactant that is used up when a reaction goes to completion even though other reactants are not completely consumed.

8.11 The main difference is that when stoichiometric quantities are used, every reactant is used up when the reaction is completed, whereas when a limiting reactant is used, it is used up and the other reactants will be left over when the reaction is completed.

8.12 For a given reaction, the theoretical yield is the maximum mass of a product that you can obtain from a given amounts of reactants. The actual yield is the mass of product that you actually recover. The percentage yield is the actual yield (from experiment) expressed as a percentage of the theoretical yield (from calculation), or

$$\text{percentage yield} = \frac{\text{actual yield}}{\text{theoretical yield}} \times 100\%$$

■ Solutions to Practice Problems

Note on significant figures: When the final answer is written for the first time, it is written with one nonsignificant digit, and the least significant digit is underlined. The final answer is then rounded to the correct number of significant figures.

8.13 $1.86 \text{ mol Al} \times \dfrac{3 \text{ mol O}_2}{4 \text{ mol Al}} = 1.3\underline{9}5 = 1.40 \text{ mol O}_2$

8.15 First, balance the equation.

$$C_2H_5OH(g) + 3\ O_2(g) \longrightarrow 2\ CO_2(g) + 3\ H_2O(g)$$

Thus, the moles of oxygen required to react with 8.24 mol ethanol is

$$8.24 \text{ mol } C_2H_5OH \times \dfrac{3 \text{ mol O}_2}{1 \text{ mol } C_2H_5OH} = 24.\underline{7}2 = 24.7 \text{ mol O}_2$$

8.17 $5.6 \text{ mol } I_2 \times \dfrac{2 \text{ mol AlI}_3}{3 \text{ mol } I_2} = 3.\underline{7}3 = 3.7 \text{ mol AlI}_3$

8.19 $3.00 \text{ mol NaOH} \times \dfrac{1 \text{ mol Na}_2SO_4}{2 \text{ mol NaOH}} = 1.5\underline{0}0 = 1.50 \text{ mol Na}_2SO_4$

8.21 $3.56 \text{ mol HNO}_3 \times \dfrac{3 \text{ mol NO}_2}{2 \text{ mol HNO}_3} = 5.3\underline{4}0 = 5.34 \text{ mol NO}_2$

8.23 The moles of sodium carbonate required are

$$1.60 \text{ mol NaHCO}_3 \times \frac{1 \text{ mol Na}_2\text{CO}_3}{2 \text{ mol NaHCO}_3} = 0.800\underline{0} = 0.800 \text{ mol Na}_2\text{CO}_3$$

For carbon dioxide, the required amount is

$$1.60 \text{ mol NaHCO}_3 \times \frac{1 \text{ mol CO}_2}{2 \text{ mol NaHCO}_3} = 0.800\underline{0} = 0.800 \text{ mol CO}_2$$

8.25 The molar masses of sodium (Na) and chlorine (Cl_2) are 22.99 g and 70.90 g, respectively. Thus,

$$1.00 \text{ g Cl}_2 \times \frac{1 \text{ mol Cl}_2}{70.90 \text{ g Cl}_2} \times \frac{2 \text{ mol Na}}{1 \text{ mol Cl}_2} \times \frac{22.99 \text{ g Na}}{1 \text{ mol Na}}$$

$$= 0.64\underline{85} = 0.649 \text{ g Na}$$

8.27 The molar masses of chromium (Cr) and hydrochloric acid (HCl) are 52.00 g and 36.46 g, respectively. Thus,

$$1.00 \text{ g Cr} \times \frac{1 \text{ mol Cr}}{52.00 \text{ g Cr}} \times \frac{2 \text{ mol HCl}}{1 \text{ mol Cr}} \times \frac{36.46 \text{ g HCl}}{1 \text{ mol HCl}}$$

$$= 1.4\underline{02} = 1.40 \text{ g HCl}$$

8.29 The molar masses of calcium carbonate ($CaCO_3$) and calcium oxide (CaO) are 100.09 g and 56.08 g, respectively. Thus,

$$1.62 \text{ g CaCO}_3 \times \frac{1 \text{ mol CaCO}_3}{100.09 \text{ g CaCO}_3} \times \frac{1 \text{ mol CaO}}{1 \text{ mol CaCO}_3} \times \frac{56.08 \text{ g CaO}}{1 \text{ mol CaO}}$$

$$= 0.90\underline{77} = 0.908 \text{ g CaO}$$

QUANTITIES IN CHEMICAL REACTIONS   81

8.31 The molar masses of mercury(II) oxide (HgO) and mercury (Hg) are 216.59 g and 200.59 g, respectively. Thus,

$$10.3 \text{ g HgO} \times \frac{1 \text{ mol HgO}}{216.59 \text{ g HgO}} \times \frac{2 \text{ mol Hg}}{2 \text{ mol HgO}} \times \frac{200.59 \text{ g Hg}}{1 \text{ mol Hg}}$$

$$= 9.5\underline{3}9 = 9.54 \text{ g Hg}$$

8.33 The molar masses of hydrochloric acid (HCl) and ammonium chloride (NH$_4$Cl) are 36.46 g and 53.49 g, respectively. Thus,

$$2.36 \text{ g NH}_4\text{Cl} \times \frac{1 \text{ mol NH}_4\text{Cl}}{53.49 \text{ g NH}_4\text{Cl}} \times \frac{1 \text{ mol HCl}}{1 \text{ mol NH}_4\text{Cl}} \times \frac{36.46 \text{ g HCl}}{1 \text{ mol HCl}}$$

$$= 1.6\underline{0}9 = 1.61 \text{ g HCl}$$

8.35 The molar masses of sodium (Na) and hydrogen (H$_2$) are 22.99 g and 2.016 g, respectively. Thus,

$$3.5 \text{ g H}_2 \times \frac{1 \text{ mol H}_2}{2.016 \text{ g H}_2} \times \frac{2 \text{ mol Na}}{1 \text{ mol H}_2} \times \frac{22.99 \text{ g Na}}{1 \text{ mol Na}} = 7\underline{9}.8 = 80. \text{ g Na}$$

8.37 In each part, calculate the required amount of water necessary to react with the given amount of KO$_2$.

a. $6.4 \text{ mol KO}_2 \times \dfrac{2 \text{ mol H}_2\text{O}}{4 \text{ mol KO}_2} = 3.2 \text{ mol H}_2\text{O}$

Since the available amount of water (2.1 mol) is less than the required amount, H$_2$O is the limiting reactant.

b. $8.4 \text{ mol KO}_2 \times \dfrac{2 \text{ mol H}_2\text{O}}{4 \text{ mol KO}_2} = 4.2 \text{ mol H}_2\text{O}$

Since the available amount of water (1.5 mol) is less than the required amount, H$_2$O is the limiting reactant.

c. $8.4 \text{ mol } KO_2 \times \dfrac{2 \text{ mol } H_2O}{4 \text{ mol } KO_2} = 4.2 \text{ mol } H_2O$

Since the available amount of water (2.1 mol) is less than the required amount, H_2O is the limiting reactant.

8.39 a. The balanced equation is

$$P_4(s) + 6 Cl_2(g) \longrightarrow 4 PCl_3(s)$$

The amount of Cl_2 required to react with 2.00 mol P_4 is

$$2.0 \text{ mol } P_4 \times \dfrac{6 \text{ mol } Cl_2}{1 \text{ mol } P_4} = 12 \text{ mol } Cl_2$$

Since the given amount of Cl_2 (2.00 mol) is less than the required amount, Cl_2 is the limiting reactant.

b. The balanced equation is

$$2 Al(s) + 3 Cl_2(g) \longrightarrow 2 AlCl_3(s)$$

The amount of Cl_2 required to react with 2.0 mol Al is

$$2.0 \text{ mol Al} \times \dfrac{3 \text{ mol } Cl_2}{2 \text{ mol Al}} = 3.0 \text{ mol } Cl_2$$

Since the given amount of Cl_2 (2.0 mol) is less than the required amount, Cl_2 is the limiting reactant.

c. The balanced equation is

$$C(s) + 2 Cl_2(g) \longrightarrow CCl_4(l)$$

The amount of Cl_2 required to react with 2.0 mol C is

$$2.0 \text{ mol C} \times \dfrac{2 \text{ mol } Cl_2}{1 \text{ mol C}} = 4.0 \text{ mol } Cl_2$$

Since the given amount of Cl_2 (2.0 mol) is less than the required amount, Cl_2 is the limiting reactant.

QUANTITIES IN CHEMICAL REACTIONS

8.41 First, determine the limiting reactant, if there is one. Calculate the required amount of Cl_2 necessary to react with 3.0 mol Na.

$$3.0 \text{ mol Na} \times \frac{1 \text{ mol } Cl_2}{2 \text{ mol Na}} = 1.5 \text{ mol } Cl_2$$

Since the given amount of Cl_2 (3.0 mol) is more than the required amount, Na is the limiting reactant, and will be used up in the reaction. Next, calculate the amount of NaCl that will be formed.

$$3.0 \text{ mol Na} \times \frac{2 \text{ mol NaCl}}{2 \text{ mol Na}} = 3.0 \text{ mol NaCl}$$

Finally, determine the amount of Cl_2 that will remain after the reaction is over. The amount used up in the reaction is 1.5 mol. Thus,

$$3.0 \text{ mol } Cl_2 - 1.5 \text{ mol } Cl_2 = 1.5 \text{ mol } Cl_2$$

8.43 First, determine the limiting reactant, if there is one. Calculate the number of moles of Al required to react with 2.5 mol O_2.

$$2.5 \text{ mol } O_2 \times \frac{4 \text{ mol Al}}{3 \text{ mol } O_2} = 3.33 \text{ mol Al}$$

Since the available amount of Al (2.5 mol) is less than the required amount, Al is the limiting reactant, and is used up in the reaction. Next, use the limiting reactant to calculate the amount of Al_2O_3 that is formed.

$$2.5 \text{ mol Al} \times \frac{2 \text{ mol } Al_2O_3}{4 \text{ mol Al}} = 1.2\underline{5} = 1.3 \text{ mol } Al_2O_3$$

Finally, determine the amount of O_2 that will remain after the reaction is over. The amount used up in the reaction is

$$2.5 \text{ mol Al} \times \frac{3 \text{ mol } O_2}{4 \text{ mol Al}} = 1.88 \text{ mol } O_2$$

Thus, the amount remaining is

$$2.5 \text{ mol } O_2 - 1.88 \text{ mol } O_2 = 0.62 \text{ mol } O_2$$

8.45 First, convert the given mass to moles. The molar masses of NH_3 and HCl are 17.03 g and 36.46 g, respectively. Thus,

$$1.00 \text{ g } NH_3 \times \frac{1 \text{ mol } NH_3}{17.03 \text{ g } NH_3} = 0.05872 \text{ mol } NH_3$$

$$1.00 \text{ g HCl} \times \frac{1 \text{ mol HCl}}{36.46 \text{ g HCl}} = 0.02743 \text{ mol HCl}$$

Next, determine the limiting reactant, if there is one. Calculate the amount of HCl required to react with 0.05872 mol NH_3.

$$0.05872 \text{ mol } NH_3 \times \frac{1 \text{ mol HCl}}{1 \text{ mol } NH_3} = 0.05872 \text{ mol HCl}$$

Since the available amount of HCl (0.02743 mol) is less than the required amount, HCl is the limiting reactant. Use the limiting reactant to determine the amount of NH_3 that will be used up in the reaction.

$$0.02743 \text{ mol HCl} \times \frac{1 \text{ mol } NH_3}{1 \text{ mol HCl}} \times \frac{17.03 \text{ g } NH_3}{1 \text{ mol } NH_3} = 0.4671 \text{ g } NH_3$$

Thus, the amount of NH_3 present at the end of the reaction is

$$1.00 \text{ g } NH_3 - 0.467 \text{ g } NH_3 = 0.53\underline{3} = 0.53 \text{ g } NH_3$$

8.47 First, convert the given masses to moles. The molar masses of CO and H_2 are 28.01 g and 2.016 g, respectively. Thus,

$$30.0 \text{ g CO} \times \frac{1 \text{ mol CO}}{28.01 \text{ g CO}} = 1.071 \text{ mol CO}$$

$$30.0 \text{ g } H_2 \times \frac{1 \text{ mol } H_2}{2.016 \text{ g } H_2} = 14.88 \text{ mol } H_2$$

(continued)

Next, determine the limiting reactant, if there is one. Calculate the amount of H_2 required to react with 1.071 mol CO.

$$1.071 \text{ mol CO} \times \frac{2 \text{ mol } H_2}{1 \text{ mol CO}} = 2.142 \text{ mol } H_2$$

Since the available amount of H_2 (14.88 mol) is more than the required amount, CO is the limiting reactant. Next, use the limiting reactant to calculate the mass of CH_3OH that will be formed and the mass of H_2 that will react. The molar mass of CH_3OH is 32.04 g.

$$1.071 \text{ mol CO} \times \frac{1 \text{ mol } CH_3OH}{1 \text{ mol CO}} \times \frac{32.04 \text{ g } CH_3OH}{1 \text{ mol } CH_3OH}$$

$$= 34.\underline{3}1 = 34.3 \text{ g } CH_3OH$$

$$1.071 \text{ mol CO} \times \frac{2 \text{ mol } H_2}{1 \text{ mol CO}} \times \frac{2.016 \text{ g } H_2}{1 \text{ mol } H_2} = 4.3\underline{1}8 \text{ g } H_2$$

Finally, determine the mass of H_2 remaining when the reaction is over.

$$30.0 \text{ g } H_2 - 4.318 \text{ g } H_2 = 25.\underline{6}8 = 25.7 \text{ g } H_2$$

8.49 percentage yield = $\dfrac{4.72 \text{ g}}{4.87 \text{ g}} \times 100\%$ = 96.$\underline{9}$2 = 96.9%

8.51 First, calculate the theoretical yield. The molar mass of nitrogen dioxide, NO_2, is 46.01 g, and for HNO_3 it is 63.02 g.

$$60.0 \text{ g } NO_2 \times \frac{1 \text{ mol } NO_2}{46.01 \text{ g } NO_2} \times \frac{2 \text{ mol } HNO_3}{3 \text{ mol } NO_2} \times \frac{63.02 \text{ g } HNO_3}{1 \text{ mol } HNO_3}$$

$$= 54.\underline{7}9 = 54.8 \text{ g } HNO_3$$

Thus, the percentage yield of HNO_3 is

percentage yield = $\dfrac{44.2 \text{ g}}{54.79 \text{ g}} \times 100\%$ = 80.$\underline{6}$7 = 80.7%

86 ■ CHAPTER 8

■ Solutions to Additional Problems

8.53 The balanced equation is

$$CaCO_3 + 2\,HCl \longrightarrow CaCl_2 + CO_2 + H_2O$$

The moles of HCl required is

$$22.4 \text{ mol CaCO}_3 \times \frac{2 \text{ mol HCl}}{1 \text{ mol CaCO}_3} = 44.8 \text{ mol HCl}$$

The moles formed are

$$22.4 \text{ mol CaCO}_3 \times \frac{1 \text{ mol CaCl}_2}{1 \text{ mol CaCO}_3} = 22.4 \text{ mol CaCl}_2$$

$$22.4 \text{ mol CaCO}_3 \times \frac{1 \text{ mol CO}_2}{1 \text{ mol CaCO}_3} = 22.4 \text{ mol CO}_2$$

$$22.4 \text{ mol CaCO}_3 \times \frac{1 \text{ mol H}_2\text{O}}{1 \text{ mol CaCO}_3} = 22.4 \text{ mol H}_2\text{O}$$

8.55 The balanced equation is

$$Fe_2O_3 + 6\,HCl \longrightarrow 2\,FeCl_3 + 3\,H_2O$$

The moles of HCl required is

$$1.23 \text{ mol Fe}_2\text{O}_3 \times \frac{6 \text{ mol HCl}}{1 \text{ mol Fe}_2\text{O}_3} = 7.38 \text{ mol HCl}$$

(continued)

The moles formed are

$$1.23 \text{ mol Fe}_2\text{O}_3 \times \frac{2 \text{ mol FeCl}_3}{1 \text{ mol Fe}_2\text{O}_3} = 2.46 \text{ mol FeCl}_3$$

$$1.23 \text{ mol Fe}_2\text{O}_3 \times \frac{3 \text{ mol H}_2\text{O}}{1 \text{ mol Fe}_2\text{O}_3} = 3.69 \text{ mol H}_2\text{O}$$

8.57 The balanced equation is

$$3 \text{ NaOH} + \text{H}_3\text{PO}_4 \longrightarrow \text{Na}_3\text{PO}_4 + 3 \text{ H}_2\text{O}$$

Convert the masses to moles. The molar masses of NaOH and H_3PO_4 are 40.00 g and 97.99 g, respectively.

$$5.00 \text{ g NaOH} \times \frac{1 \text{ mol NaOH}}{40.00 \text{ g NaOH}} = 0.1250 \text{ mol NaOH}$$

Determine the limiting reactant, if there is one. Calculate the amount of H_3PO_4 required to react with 0.1250 mol NaOH.

$$7.00 \text{ g H}_3\text{PO}_4 \times \frac{1 \text{ mol H}_3\text{PO}_4}{97.99 \text{ g H}_3\text{PO}_4} = 0.07144 \text{ mol H}_3\text{PO}_4$$

$$0.1250 \text{ mol NaOH} \times \frac{1 \text{ mol H}_3\text{PO}_4}{3 \text{ mol NaOH}} = 0.04167 \text{ mol H}_3\text{PO}_4$$

Since the available H_3PO_4 (0.07144 mol) is more than the required amount, NaOH is the limiting reactant. The molar mass of Na_3PO_4 is 163.94 g. Therefore, the mass of Na_3PO_4 formed is

$$0.1250 \text{ mol NaOH} \times \frac{1 \text{ mol Na}_3\text{PO}_4}{3 \text{ mol NaOH}} \times \frac{163.94 \text{ g Na}_3\text{PO}_4}{1 \text{ mol Na}_3\text{PO}_4}$$

$$= 6.8\underline{3}1 = 6.83 \text{ g Na}_3\text{PO}_4$$

(continued)

The mass of H_3PO_4 that reacts is

$$0.1250 \text{ mol NaOH} \times \frac{1 \text{ mol } H_3PO_4}{3 \text{ mol NaOH}} \times \frac{97.99 \text{ g } H_3PO_4}{1 \text{ mol } H_3PO_4} = 4.083 \text{ g } H_3PO_4$$

Finally, the mass of H_3PO_4 remaining at the end of the reaction is

$$7.00 \text{ g } H_3PO_4 - 4.083 \text{ g } H_3PO_4 = 2.9\underline{1}7 = 2.92 \text{ g } H_3PO_4$$

8.59 The balanced equation is

$$Ca(HCO_3)_2 + Ca(OH)_2 \longrightarrow 2 \, CaCO_3 + 2 \, H_2O$$

The molar masses of $CaCO_3$ and $Ca(HCO_3)_2$ are 100.09 g and 162.12 g, respectively. Thus, the mass of $CaCO_3$ formed is

$$16.3 \text{ g } Ca(HCO_3)_2 \times \frac{1 \text{ mol } Ca(HCO_3)_2}{162.12 \text{ g } Ca(HCO_3)_2} \times \frac{2 \text{ mol } CaCO_3}{1 \text{ mol } Ca(HCO_3)_2}$$

$$\times \frac{100.09 \text{ g } CaCO_3}{1 \text{ mol } CaCO_3} = 20.\underline{1}3 = 20.1 \text{ g } CaCO_3$$

8.61 The balanced equation is

$$3 \, Ba(OH)_2 + \overset{\sim}{\boxed{3}} H_3PO_4 \longrightarrow Ba_3(PO_4)_2 + 6 \, H_2O$$

The molar masses of $Ba(OH)_2$ and $Ba_3(PO_4)_2$ are 171.35 g and 601.93 g, respectively. Thus, the mass of $Ba(OH)_2$ required is

$$3.00 \text{ g } Ba_3(PO_4)_2 \times \frac{1 \text{ mol } Ba_3(PO_4)_2}{601.93 \text{ g } Ba_3(PO_4)_2} \times \frac{3 \text{ mol } Ba(OH)_2}{1 \text{ mol } Ba_3(PO_4)_2}$$

$$\times \frac{171.35 \text{ g } Ba(OH)_2}{1 \text{ mol } Ba(OH)_2} = 2.5\underline{6}2 = 2.56 \text{ g } Ba(OH)_2$$

(continued)

The molar mass of H_3PO_4 is 97.99 g. Thus, the mass of H_3PO_4 required is

$$3.00 \text{ g } Ba_3(PO_4)_2 \times \frac{1 \text{ mol } Ba_3(PO_4)_2}{601.93 \text{ g } Ba_3(PO_4)_2} \times \frac{2 \text{ mol } H_3PO_4}{1 \text{ mol } Ba_3(PO_4)_2}$$

$$\times \frac{97.99 \text{ g } H_3PO_4}{1 \text{ mol } H_3PO_4} = 0.97\underline{6}8 = 0.977 \text{ g } H_3PO_4$$

8.63 The balanced equations are

$$2 C + O_2 \longrightarrow 2 CO$$

$$ZnO + CO \longrightarrow Zn + CO_2$$

First, calculate the moles of CO that can be formed from 50.0 g of carbon. The molar mass of C is 12.01 g. Thus,

$$50.0 \text{ g} \times \frac{1 \text{ mol C}}{12.01 \text{ g C}} \times \frac{2 \text{ mol CO}}{2 \text{ mol C}} = 4.163 \text{ mol CO}$$

Next, convert the 75.0 g of zinc oxide (ZnO) to moles. The molar mass of ZnO is 81.38 g. Thus,

$$75.0 \text{ g ZnO} \times \frac{1 \text{ mol ZnO}}{81.38 \text{ g ZnO}} = 0.9216 \text{ mol ZnO}$$

Next, determine the limiting reactant, if there is one. The amount of CO required to react with 0.9216 mol ZnO is

$$0.9216 \text{ mol ZnO} \times \frac{1 \text{ mol CO}}{1 \text{ mol ZnO}} = 0.9216 \text{ mol CO}$$

Since the available CO (4.163 mol) is more than the required amount, ZnO is the limiting reactant. Since the molar mass of zinc is 65.83 g, the mass of Zn formed is

$$0.9216 \text{ mol ZnO} \times \frac{1 \text{ mol Zn}}{1 \text{ mol ZnO}} \times \frac{65.38 \text{ g Zn}}{1 \text{ mol Zn}} = 60.\underline{2}5 = 60.3 \text{ g Zn}$$

8.65 The balanced equation is

$$C_7H_6O_3 + C_4H_6O_3 \longrightarrow C_9H_8O_4 + HC_2H_3O_2$$

First, convert the masses of reactants to moles. The molar masses of $C_7H_6O_3$ and $C_4H_6O_3$ are 138.12 g and 102.09 g, respectively. Thus,

$$2.00 \text{ g } C_7H_6O_3 \times \frac{1 \text{ mol } C_7H_6O_3}{138.12 \text{ g } C_7H_6O_3} = 0.01448 \text{ mol } C_7H_6O_3$$

$$4.00 \text{ g } C_4H_6O_3 \times \frac{1 \text{ mol } C_4H_6O_3}{102.09 \text{ g } C_4H_6O_3} = 0.03918 \text{ mol } C_4H_6O_3$$

Next, determine the limiting reactant, if there is one. The amount of $C_4H_6O_3$ required to react with 0.01448 mol $C_7H_6O_3$ is

$$0.01448 \text{ mol } C_7H_6O_3 \times \frac{1 \text{ mol } C_4H_6O_3}{1 \text{ mol } C_7H_6O_3} = 0.01448 \text{ mol } C_4H_6O_3$$

Since the available $C_4H_6O_3$ (0.03918 mol) is more than the required amount, $C_7H_6O_3$ is the limiting reactant. The molar mass of aspirin ($C_9H_8O_4$) is 180.15 g. Thus, the mass of aspirin formed is

$$0.01448 \text{ mol } C_7H_6O_3 \times \frac{1 \text{ mol } C_9H_8O_4}{1 \text{ mol } C_7H_6O_3} \times \frac{180.15 \text{ g } C_9H_8O_4}{1 \text{ mol } C_9H_8O_4}$$

$$= 2.6\underline{0}9 = 2.61 \text{ g } C_9H_8O_4$$

Solutions to Practice in Problem Analysis

8.1 To solve this problem, you must first convert the mass of iron (5.0 g) to moles using the molar mass of iron (55.85 g/mol) as a conversion factor. Next, the moles of iron are converted to moles of iron(III) oxide using the mole ratio from the balanced equation (2 mol Fe_2O_3/ 4 mol Fe). Finally, convert the moles of iron(III) oxide to grams using the molar mass of iron(III) oxide as a conversion factor (159.70 g/mol).

8.2 To solve this problem, you must first determine which of the reactants, if either, is the limiting reactant. Select one of the reactants, say NaOH (5.0 g) and calculate the amount of the other reactant, H_2SO_4, that reacts completely with the NaOH. If the actual amount of H_2SO_4 is equal to the calculated amount, the reactants are present in stoichiometric amounts, and there is no limiting reactant. If the actual amount of H_2SO_4 is greater than the calculated amount, it is in excess and NaOH is the limiting reactant. If the actual amount is less than the calculated amount, H_2SO_4 is the limiting reactant, and NaOH is in excess. To do the determination, you will also need the molar masses of NaOH (40.00 g/mol) and H_2SO_4 (98.09 g/mol) as conversion factors. In addition, the mole ratio from the balanced equation (1 mol H_2SO_4/ 2 mol NaOH) is needed as a conversion factor. Next, use the available quantity of the limiting reactant to calculate the quantity of sodium sulfate that will form. You will need the molar mass of Na_2SO_4 (142.05 g/mol) as a conversion factor.

Answers to the Practice Exam

1. e   2. a   3. c   4. e   5. e   6. a   7. e   8. d   9. b   10. c

11. a   12. a   13. c   14. c   15. b

9. ELECTRON STRUCTURE OF ATOMS

■ Solutions to Exercises

9.1   There are $3^2 = 9$ orbitals in the n = 3 shell.

9.2   The notation 5d refers to the n = 5 shell and a d-type subshell.

9.3   In the n = 3 shell, there are 3 subshells. Each subshell will begin with 3. They are 3s, 3p, and 3d. The numbers of orbitals in each subshell are 1, 3, and 5, respectively.

9.4   The superscripts show that there are 2 electrons in the 1s subshell, 2 electrons in the 2s subshell, 6 electrons in the 2p subshell, 2 electrons in the 3s subshell, and 5 electrons in the 3p subshell. The total number of electrons is 2 + 2 + 6 + 2 + 5 = 17.

9.5   Chlorine has atomic number 17 and therefore 17 electrons. The subshells, in order of filling, are 1s, 2s, 2p, 3s, and 3p. Thus, the electron configuration is $1s^2 2s^2 2p^6 3s^2 3p^5$.

9.6   The subshells through germanium (atomic number 32) are 1s, 2s, 2p, 3s, 3p, 4s, 3d, and 4p. Putting the electrons into the orbitals, you get $1s^2 2s^2 2p^6 3s^2 3p^6 4s^2 3d^{10} 4p^2$. It is convenient to rearrange the subshells so that they are in order by shells. This gives the electron configuration $1s^2 2s^2 2p^6 3s^2 3p^6 3d^{10} 4s^2 4p^2$.

9.7   The row number of germanium is 4, so n = 4, and the general form of the valence-shell configuration is $4s^a 4p^b$. The group number is 4 (Roman numeral IV), so a + b = 4. The valence-shell configuration is $4s^2 4p^2$.

9.8   Since O and F are in the same period, F is the smaller atom.

9.9   Since F^- and Cl^- are in the same group (Group VIIA), F^- is smaller.

9.10 The atom with the smaller ionization energy is O.

Answers to Questions to Test Your Reading

9.1 When ordinary white light is separated into its components by means of a prism, you obtain a continuous spectrum consisting of a continuous rainbow of colors. But, when you separate the light emitted by heated atoms of an element, you see a line spectrum, or a spectrum consisting of a series of colored lines against a black background. These line spectra are characteristic of a particular kind of atom.

9.2 In Bohr's theory of the atom, it is assumed that the electrons in atoms move in orbits about their nucleus with only certain allowed energies, or energy levels. Also, electrons undergo transitions between energy levels. When electrons undergo a transition to a lower energy level, energy is released as a photon with a particular wavelength, or color.

9.3 An energy level of an atom is one of the allowed energy values that an electron can have, and is designated by a quantum number, n, which is a positive integer.

9.4 The modern theory of atoms is called quantum mechanics.

9.5 An orbit is a specific circular path about the nucleus, much like the earth moves about the sun. An orbital is the wave function for an electron in an atom. Roughly speaking, it is the region of space around the nucleus where the electron is likely to be found. It is much less well defined than an orbit; instead of moving over a definite path, the electron occupies a region of space. The electron's exact position at any given time within that region of space is not precisely predictable.

9.6 An electron shell is a set of orbitals of approximately the same size and energy. Different shells correspond to regions of space that are different distances from the nucleus. Each electron shell is designated by a shell quantum number, n. A subshell is a subset of orbitals of an electron shell, each orbital having identical energy and specific shapes. The subshells are denoted s, p, d, f, g, h, and so forth.

9.7 The different possible values of n are any positive whole number value from 1 to infinity.

9.8 For a given shell whose quantum number is n, there are n subshells. They are denoted s, p, d, f, g, h, and so forth, and are preceded by the shell quantum number. For example, a p-type subshell in the n = 2 shell is denoted 2p.

9.9 The number of orbitals in the nth shell of an atom equals n^2. For the $n = 5$ shell, there should be $5^2 = 25$ orbitals. These can be summarized as follows:

subshell	number of orbitals
5s	1
5p	3
5d	5
5f	7
5g	9
Total =	25

The total number of orbitals agrees with the predicted number, 25.

9.10 All s orbitals have a spherical shape, and p orbitals have a dumbbell shape.

9.11 The order (by energy) of the first four subshells of a given shell are $s < p < d < f$.

9.12 An electron configuration is a particular distribution of electrons among the different subshells of an atom. It is described by listing the symbols for the occupied subshells one after another, adding a superscript to each symbol to show the number of electrons in the subshell.

9.13 According to the Pauli exclusion principle, an orbital can hold no more than two electrons. Since each subshell has a specified number of orbitals, the maximum number of electrons that a subshell can accommodate is twice the number of orbitals in the subshell. For the $n = 1$ shell, there is only one subshell (1s) and only one orbital (1s). Thus, the $n = 1$ shell can accommodate $2 \times 1 = 2$ electrons. For the $n = 2$ shell, there are two subshells. The 2s subshell has one orbital, and the 2p subshell has three orbitals. Thus, the $n = 2$ shell has $1 + 3 = 4$ orbitals. Since each orbital can accommodate 2 electrons, the maximum number of electrons that the $n = 2$ shell can accommodate is $2 \times 4 = 8$ electrons.

9.14 The first shell ($n = 1$) can accommodate 2 electrons. Thus, there are two different electron configurations possible for the $n = 1$ shell. These are $1s^1$ and $1s^2$. Thus, the first row of the periodic table has two elements. The second shell can accommodate 8 electrons. Thus, the second row of the periodic table has eight elements.

9.15 The first three noble gases have the following electron configurations:

noble gas	electron configuration
helium (He)	$1s^2$
neon (Ne)	$1s^2 2s^2 2p^6$
argon (Ar)	$1s^2 2s^2 2p^6 3s^2 3p^6$

9.16 According to the periodic table, the first nine electron subshells in order of filling are: 1s, 2s, 2p, 3s, 3p, 4s, 3d, 4p, 5s.

9.17 Valence electrons are the outer electrons in an atom. They are responsible for the chemical reactivity of an atom.

9.18 Carbon is in Group 4 (Roman numeral IV). Thus, a + b = 4. The valence-shell electron configuration for a carbon atom is $2s^2 2p^2$.

9.19 First, the atomic radius increases in any given main-group as you move down the group of elements. Second, the atomic radius decreases in any given period of main-group elements as you move across the period.

9.20 Since barium is below strontium, it has an additional shell of electrons and therefore a larger atomic radius.

9.21 For an isoelectronic series of ions, the ionic radius decreases as the atomic number increases.

9.22 $K + \text{energy} \longrightarrow K^+ + e^-$

9.23 Since strontium and barium are both in Group IIA, and barium is below strontium, the ionization energy will decrease as you progress from strontium to barium. This is because the atoms are getting bigger, and the outer electrons are not held as tightly. Thus, they can be removed easier.

9.24 Electron affinity refers to how strongly an atom tends to gain electrons from other atoms.

9.25 The ionization energies of nonmetals are larger than the ionization energies of metals. The most reactive metals have the lowest ionization energies, while the most reactive nonmetals have the highest ionization energies.

9.26 The strongest electron affinities are for the elements fluorine, oxygen, and chlorine.

■ Solutions to Practice Problems

9.27 Since shorter wavelengths correspond to higher frequencies, violet light, with a wavelength of 400 nm, will have the higher frequency.

9.29 Since shorter wavelengths correspond to higher energies, green light, with a wavelength of 500 nm, will have the higher energy.

9.31 Since higher values of n correspond to higher energy, the n = 4 energy level has the greater energy.

9.33 For both hydrogen atoms, the electron starts in the same energy level (n = 5). The electron that undergoes a transition to a lower final energy state will release more energy than a transition to a higher final energy state. Thus, the transition from the n = 5 level to the n = 2 level will emit a photon with greater energy.

9.35 a. $2^2 = 4$ b. $5^2 = 25$

9.37 a. The notation 3p refers to the n = 3 shell and a p-type subshell.

b. The notation 4d refers to the n = 4 shell and a d-type subshell.

c. The notation 2s refers to the n = 2 shell and a s-type subshell.

9.39 Because n = 2, there are 2 subshells, 2s and 2p. The number of orbitals in each subshell are 1 and 3, respectively.

9.41 3s < 3p < 3d

9.43 The superscripts show that there are 2 electrons in the 1s subshell, 2 electrons in the 2s subshell, 6 electrons in the 2p subshell, 2 electrons in the 3s subshell, and 4 electrons in the 3p subshell. The total number of electrons in the atom is 2 + 2 + 6 + 2 + 4 = 16.

9.45 The n = 2 shell can hold 2×2^2 = 8 electrons.

9.47 The electron configuration for a silicon atom is $1s^22s^22p^63s^23p^2$.

9.49 The electron configuration for an arsenic atom is $1s^22s^22p^63s^23p^63d^{10}4s^24p^3$.

9.51 Strontium (Sr) is in Group IIA (or 2) and period 5. Thus, the valence-shell configuration is $5s^2$.

9.53 The valence-shell configuration for this atom is $4s^24p^4$. Thus, it is in period 4 and Group VIA. The element is selenium (Se).

9.55 a. Br is larger than F.   b. K is larger than Br.

9.57 a. O^{2-} is smaller than Se^{2-}.   b. Cl^- is smaller than S^{2-}.

9.59 a. F has a greater ionization energy than Br.

b. Br has a greater ionization energy than K.

ELECTRON STRUCTURE OF ATOMS 97

Solutions to Additional Problems

9.61 a. 13   b. 5   c. 18   d. 35

9.63 a. 2p is allowed.

b. 2d is not allowed. The n = 2 shell has only 2 subshells, 2s and 2p.

c. 3f is not allowed. The n = 3 shell has only 3 subshells, 3s, 3p, and 3d.

d. 4f is allowed.

9.65 a. lowest energy

b. lowest energy

c. not allowed

d. lowest energy

9.67 The first four elements of Group IVA have the following electron configurations:

element	electron configuration
carbon (C)	$1s^2 2s^2 2p^2$
silicon (Si)	$1s^2 2s^2 2p^6 3s^2 3p^2$
germanium (Ge)	$1s^2 2s^2 2p^6 3s^2 3p^6 3d^{10} 4s^2 4p^2$
tin (Sn)	$1s^2 2s^2 2p^6 3s^2 3p^6 3d^{10} 4s^2 4p^6 4d^{10} 5s^2 5p^2$

9.69 The total number of electrons in the atom, and thus the atomic number, is 2 + 2 + 6 + 2 + 2 + = 22. This corresponds to the element titanium (Ti).

9.71 a. 1   b. 2   c. 4   d. 6

9.73 The first four elements of Group IIIA have the following valence-shell configurations:

element	valence-shell configuration
boron (B)	$2s^2 2p^1$
aluminum (Al)	$3s^2 3p^1$
gallium (Ga)	$4s^2 4p^1$
indium	$5s^2 5p^1$

9.75 Be

9.77 Li is the largest atom in period 2.

9.79 In order of increasing atomic radius: F < Cl < S.

9.81 In order of decreasing ionization energy: S > Mg > Sr.

9.83 In order of increasing size: $F^- < Cl^- < Br^- < I^-$.

9.85 In order of decreasing size: $Se^{2-} > Br^- > Rb^+ > Sr^{2+}$.

9.87 Na is larger than Na^+.

9.89 Cl is smaller than Cl^-.

■ Solutions to Practice in Problem Analysis

9.1 Bromine (Br) is in the fourth period of Group VIIA. Looking at the pattern of the periodic table, write down the subshells in the order in which they fill (1s, 2s, 2p, 3s, 3p, 4s, 3d, 4p). Use the periodic table to note the atomic number of bromine (35) and therefore the total number of electrons in the atom. Then distribute these electrons to the subshells in the order in which they fill. Remember that an s subshell can hold only two electrons, whereas a p subshell can hold six electrons, and a d subshell can hold ten electrons. After the electrons have been placed into the subshells, it is convenient to rearrange the subshells so that they are in order by shells.

9.2 From the position of the element on the periodic table you know that the row number equals n, and the group number equals a + b. Distribute the valence electrons to the subshells, filling in the ns subshell first. If there are electrons left over, put them into the np subshell. The row number of arsenic is 4, so n = 4, and the general form of the valence shell is $4s^a 4p^b$. The group number is 5 (the Roman numeral V), so a + b = 5. Put two of these electrons in the 4s subshell and three in the 4p subshell.

■ Answers to the Practice Exam

1. e 2. b 3. d 4. b 5. c 6. b 7. a 8. e 9. e 10. d

10. CHEMICAL BONDING

■ Solutions to Exercises

10.1 a. The Al atom is the metal and is in group IIIA. The number of electrons lost by Al is 3, equal to the group number. The F atom is the nonmetal and is in Group VIIA. The number of electrons picked up by F is 8 - 7 = 1. Therefore the ions formed are Al^{3+} and F^-. Since the Al atom loses three electrons and the F atoms pick up one electron, you need one Al atom and three F atoms. Using electron-dot symbols, the equation for the formation of ions is

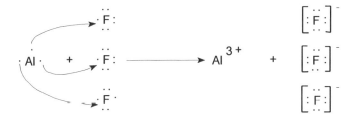

Each of the resulting ions has a noble-gas electron configuration.

b. Calcium is a metal and is in Group IIA. Thus it loses two electrons and forms the Ca^{2+} cation. Oxygen is a nonmetal and is in Group VIA. Thus it picks up 8 - 6 = 2 electrons and forms the O^{2-} anion. In simplest whole numbers the ratio of Ca^{2+} to O^{2-} is 1 to 1. The resulting equation for the formation of ions is

The resulting ions have a noble-gas electron configuration.

10.2 a. The oxygen atom in water, H_2O, is the central atom. The arrangement of the atoms in the molecule is

H O H

Step 1. The total number of valence electrons is obtained as follows: O is in Group VIA, so has 6 valence electrons; each H has 1 valence electron. So the total number of valence electrons is $1(6) + 2(1) = 8$.

Step 2. Connect the atoms with a pair of dots.

H : O : H

Step 3. Counting up the electrons used so far you get 4, which leaves $8 - 4 = 4$ electrons remaining (2 electron pairs).

Step 4. Place the electron pairs on the central atom.

H : Ö : H

Each atom satisfies the octet rule. Therefore the final electron-dot formula is

H : Ö : H   or   H — Ö — H

b. $HClO_3$, chloric acid, is an oxyacid. You place the Cl atom in the center with the O atoms around it and you bond the H atom to one of the O atoms.

```
      O
  O  Cl  O  H
```

Step 1. Hydrogen has 1 valence electron, oxygen has 6 valence electrons and chlorine has 7. The total number of valence electrons is $1(1) + 3(6) + 1(7) = 26$.

Step 2. Connect the atoms by electron pairs

```
        O
        ··
    O : Cl : O : H
```

(continued)

CHEMICAL BONDING ■ 101

Step 3. Distribute electron pairs to the O atoms to satisfy the octet rule.

$$:\!\ddot{\text{O}}\!:$$
$$:\!\ddot{\text{O}}\!:\!\ddot{\text{Cl}}\!:\!\ddot{\text{O}}\!:\text{H}$$

24 electrons have been used so far so there are 26 - 24 = 2 electrons remaining (1 electron pair).

Step 4. Place the electron pair on the central atom.

$$:\!\ddot{\text{O}}\!:$$
$$:\!\ddot{\text{O}}\!:\!\ddot{\text{Cl}}\!:\!\ddot{\text{O}}\!:\text{H}$$

Each atom satisfies the octet rule. Therefore the final electron-dot formula is

$$:\!\ddot{\text{O}}\!:$$ $$:\!\ddot{\text{O}}\!:$$
$$:\!\ddot{\text{O}}\!:\!\ddot{\text{Cl}}\!:\!\ddot{\text{O}}\!:\text{H}$$ or $$:\!\ddot{\text{O}}\!-\!\ddot{\text{Cl}}\!-\!\ddot{\text{O}}\!-\text{H}$$

10.3 a. Nitrogen, N_2, is a diatomic molecule. The arrangement of atoms is

N   N

Step 1. Each N atom has 5 valence electrons (Group VA) for a total of 2(5) = 10.

Step 2. Connect the atoms with an electron pair.

N : N

Step 3. Add electron pairs to one N atom to make an octet. Place the remaining electron pair on the other N atom.

$$:\!\ddot{\text{N}}\!:\!\text{N}\!:$$

This uses all 10 electrons available.

(continued)

Step 4. Note that the second N needs two electron pairs to satisfy its octet. Move these electron pairs from the other N into the bonding region to form a triple bond.

$$:N:::N:\quad\text{or}\quad :N\equiv N:$$

Note that all atoms have octets.

b. The carbon atom in carbon disulfide, CS_2, is more electropositive than the S atoms so it is the central atom. The arrangement of atoms is

$$S\quad C\quad S$$

Step 1. C has 4 valence electrons (Group IVA) and sulfur has 6 valence electrons (Group VIA). The total number of valence electrons is $1(4) + 2(6) = 16$.

Step 2. Connect the atoms by electron pairs.

$$S:C:S$$

Step 3. Add electron pairs to the S atoms to give them an octet.

$$:\overset{..}{S}:C:\overset{..}{S}:$$
$$\underset{..}{}\underset{..}{}$$

This uses all 16 electrons available.

Step 4. Note that the C atom requires two more electron pairs to have an octet. Move one pair of electrons from each S atom to the bonding regions, giving two double bonds.

$$\overset{..}{S}::C::\overset{..}{S}\quad\text{or}\quad \overset{..}{\underset{..}{S}}=C=\overset{..}{\underset{..}{S}}$$

Note that all atoms have octets.

10.4 The boron atom in the tetrafluoroborate ion, BF_4^-, is more electropositive than the F atoms so it is the central atom. The arrangement of atoms is

```
        F
    F   B   F
        F
```

Step 1. B has 3 valence electrons (Group IIIA) and F has 7 valence electrons (Group VIIA). You need to add one more electron to account for the negative charge on BF_4^-. Therefore, the total number of valence electrons is $1(3) + 4(7) + 1 = 32$.

Step 2. Connect the atoms by electron pairs.

```
        F
        ..
    F : B : F
        ..
        F
```

Step 3. Distribute the electron pairs around the F atoms to give them octets. This uses 32 electrons, which is all that is available.

```
         ..
       : F :
    .. .. ..
  : F : B : F :
    .. .. ..
       : F :
         ..
```

Step 4. Note that all atoms have octets. Therefore, the electron-dot formula for the ion is

$$\begin{bmatrix} & : \overset{..}{F} : & \\ : \overset{..}{\underset{..}{F}} : & B : \overset{..}{\underset{..}{F}} : \\ & : \underset{..}{F} : & \end{bmatrix}^- \quad \text{or} \quad \begin{bmatrix} & : \overset{..}{F} : & \\ : \overset{..}{\underset{..}{F}} - & B - \overset{..}{\underset{..}{F}} : \\ & : \underset{..}{F} : & \end{bmatrix}^-$$

10.5 a. The electron-dot formula of the ammonium ion, NH_4^+, is

$$\left[\begin{array}{c} H \\ \ddots \\ H : N : H \\ \ddots \\ H \end{array} \right]^+$$

Note that there are four electron pairs about the N atom. These four pairs have a tetrahedral arrangement. Thus the ion has a tetrahedral shape.

b. The electron-dot formula of nitrogen trifluoride, NF_3, is

$$: \ddot{F} : \ddot{N} : \ddot{F} :$$
$$: \ddot{F} :$$

There are four electron pairs about the N atom, and these four pairs have a tetrahedral arrangement. Looking at just the atoms and bonds, you would predict that the molecule has a trigonal pyramidal shape.

■ Answers to Questions to Test Your Reading

10.1 The two principal types of chemical bonds are covalent and ionic.

10.2 An ionic bond is the strong attractive force that exists between a positive ion and a negative ion in an ionic compound.

10.3 Positive ions and negative ions attract one another to form electrically neutral aggregates of ions, or ionic crystals. Energy is released in the process.

10.4 The lithium atom has one valence electron, which is easily lost. The ionization process can be described as follows:

Li atom $(1s^2 2s^1)$ + ionization energy $\longrightarrow$ Li^+ ion $(1s^2)$ + e^-

The resulting lithium ion has the same electron configuration as the noble gas helium.

(continued)

CHEMICAL BONDING ■ 105

The fluorine atom has seven valence electrons. It tends to gain an electron to form the fluoride ion, F⁻, releasing energy in the process. The ionization process can be described as follows:

F atom ($1s^2 2s^2 2p^5$) + e⁻ ⟶ F⁻ ion ($1s^2 2s^2 2p^6$) + energy.

The added electron goes into the 2p subshell, giving the fluoride ion an electron configuration the same as that of the noble gas neon.

10.5 The energy term that is associated with the formation of cations from atoms is the ionization energy. The corresponding term for the formation of anions from atoms is the electron affinity.

10.6 Ionic bonds are formed between atoms of metal elements with atoms of nonmetal elements.

10.7 For a main-group element, a metal forms a cation, with a charge equal to the group number. Nonmetal atoms form anions with charge equal to the group number minus 8.

10.8 a. K and O should form an ionic bond. (metal + nonmetal)

    b. C and O should form a covalent bond. (two nonmetals)

    c. Ca and Cl should form an ionic bond. (metal + nonmetal)

    d. N and F should form a covalent bond. (two nonmetals)

10.9 a. Cs will form the Cs⁺ ion. (atoms in Group IA have one valence electron)

    b. Ba will form the Ba²⁺ ion. (atoms in Group IIA have two valence electrons)

    c. S will form the S²⁻ ion. (atoms in Group VIA will gain 8 - 6 = 2 electrons)

    d. I will form the I⁻ ion. (atoms in Group VIIA will gain 8 - 7 = 1 electron)

10.10 The octet rule says that atoms tend to lose or gain electrons when bonding to give eight electrons in their valence shells. (Lithium and beryllium atoms can lose electrons and hydrogen can gain electrons to give, in each case, a single shell of two electrons, as in the noble gas helium).

10.11 a. Rb·   **b.** ·Ba·   **c.** :S̈e·   **d.** :Ï·

106 ■ CHAPTER 10

10.12 A covalent bond is a chemical bond formed by the sharing of a pair of electrons between two atoms. Often, this sharing of electrons results in an octet of electrons in the valence shells of the atoms. It is different from an ionic bond in that the atoms do not lose or gain electrons in achieving their octets. Instead, a molecular orbital forms in which the electrons occupy the region around both atoms.

10.13 An electron dot symbol is a symbol of an atom or ion. An electron dot formula represents covalent bonds between atoms in a molecule.

10.14 :Cl· + ·Cl: ⟶ :Cl:Cl: or   :Cl—Cl:

10.15a. single bond: A· + ·B ⟶ A:B

  b. double bond: :O· + ·O: ⟶ O::O   or   O=O

  c. Lone pair

10.16 A multiple bond is a double or triple bond that is formed by sharing two pairs or three pairs of electrons, respectively, between atoms in a covalent bond.

10.17 Electronegativity is a measure of the ability of an atom in a covalent bond to draw bonding electrons to itself.

10.18 The basic trends in electronegativity are first, it increases in any row (period) from left to right and second, it decreases in any column (group) from top to bottom.

10.19a. As (same period)   **b.** S (same group)
   c. Br (same period)   **d.** F (same group)

10.20
a. $\overset{\delta^+}{C}\text{—}\overset{\delta^-}{O}$ b. $\overset{\delta^+}{H}\text{—}\overset{\delta^-}{S}$ c. $\overset{\delta^-}{Cl}\text{—}\overset{\delta^+}{C}$ d. $\overset{\delta^-}{F}\text{—}\overset{\delta^+}{N}$

CHEMICAL BONDING ■ 107

10.21 a. In H_2Te, Te is the central atom since hydrogen cannot be a central atom.

b. In CCl_4, the carbon atom is more electropositive than chlorine so carbon is the central atom.

c. In OF_2, the oxygen atom is more electropositive than fluorine so oxygen is the central atom.

d. HOCl is an oxyacid. The oxygen atom is bonded to the chlorine and the hydrogen atom is bonded to the oxygen. This makes oxygen the central atom. Note that in all oxyacids with more than one oxygen atom, the atom that is not hydrogen or oxygen is the central atom.

10.22 a. In BF_4^-, boron (Group IIIA) has 3 valence electrons and fluorine (Group VIIA) has 7. You must add one electron for the negative charge on the ion. The total number of valence electrons is 1(3) + 4(7) + 1 = 32.

b. In SCl_2, sulfur (Group VIA) has 6 valence electrons and chlorine (Group VIIA) has 7. The total number of valence electrons is 1(6) + 2(7) = 20.

c. In H_2SO_4, hydrogen (Group IA) has 1 valence electron, sulfur (Group VIA) has 6, and oxygen (Group VIA) has 6. The total number of valence electrons is 2(1) + 1(6) + 4(6) = 32.

d. In PCl_3, phosphorus (Group VA) has 5 valence electrons and chlorine (Group VII) has 7. The total number of valence electrons is 1(5) + 3(7) = 26.

10.23 Bond length is the normal distance between nuclei whose atoms form a bond in a molecule.

10.24 To describe molecular structure you must specify the bond lengths and bond angles.

10.25 VSEPR, or valence shell electron pair repulsion, is a model for predicting the shapes of molecules and ions, in which the valence-shell electron pairs are arranged about each atom in such a way as to keep electron pairs as far away from one another as possible.

10.26 A molecule will be polar if, first, the component elements have a difference in electronegativities and, second, the central atom has lone pairs of electrons.

108 ■ CHAPTER 10

■ Solutions to Practice Problems

10.27a. ·Ca· + :S: ⟶ Ca²⁺ + [:S:]²⁻

b. Na· + ·O: + Na· ⟶ Na⁺ + Na⁺ + [:O:]²⁻

c. K· + ·Cl: ⟶ K⁺ + [:Cl:]⁻

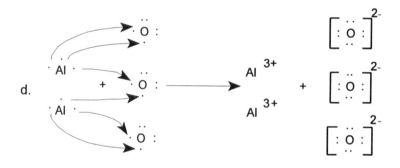

10.29a. :N≡N:

b. 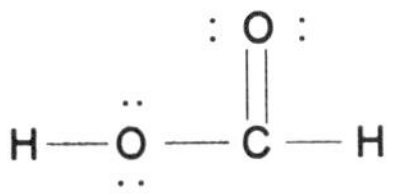

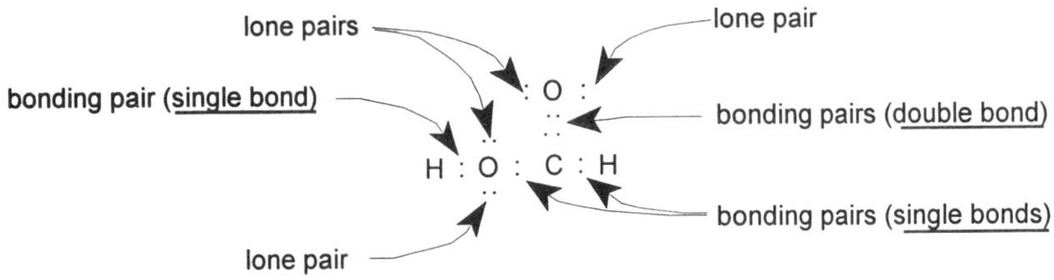

10.31a. 4.0 - 0.9 = 3.1 b. 3.0 - 2.5 = 0.5 c. 4.0 - 2.5 = 1.5 d. 2.5 - 2.5 = 0

Order of bonds (pure covalent to ionic): C–C, C–Cl, F–C, F–Na

10.33 a. 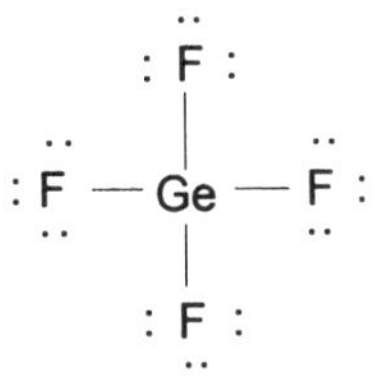 b. 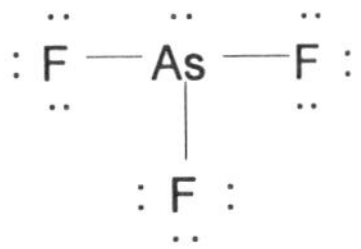

c. 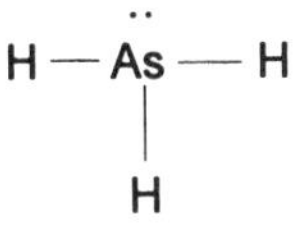 d. 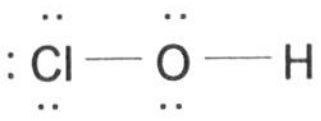

10.35 a. 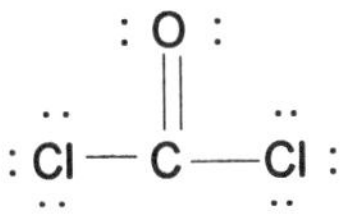 b. 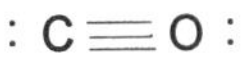 c. 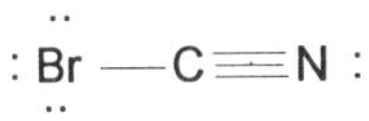

10.37 a. 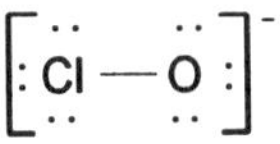 b. 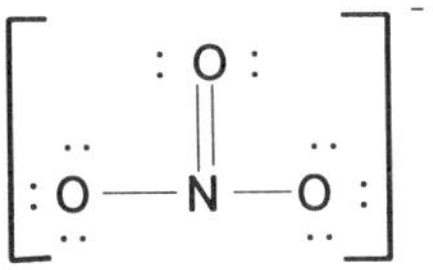

10.39 a. Tetrahedral shape

b. Trigonal pyramidal shape

c. Trigonal planar shape

Solutions to Additional Problems

10.41 LiCl is ionic (metal + nonmetal) and Cl_2 is covalent (two nonmetals). The formation of the ionic bond in LiCl can be described as follows:

Li· + ·Cl̈: ⟶ Li$^+$ + [:C̈l̈:]$^-$

The formation of the covalent bond in Cl_2 is as follows:

:C̈l· + ·C̈l: ⟶ :C̈l:C̈l: or   :C̈l—C̈l:

10.43 The electron configuration of the Pb^{2+} ion is $[Xe]4f^{14}5d^{10}6s^2$.

10.45 a. Na–O   b. H–O   c. F–H

10.47 a.   Na$^+$ + [:B̈r̈:]$^-$

b.   Ca^{2+} + [:F̈:]$^-$
                [:F̈:]$^-$

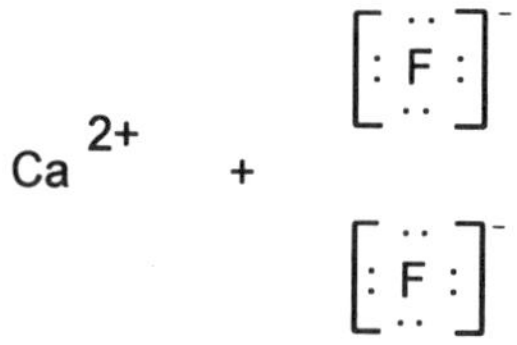

c.   Na$^+$
     Na$^+$ + [:S̈:]$^{2-}$

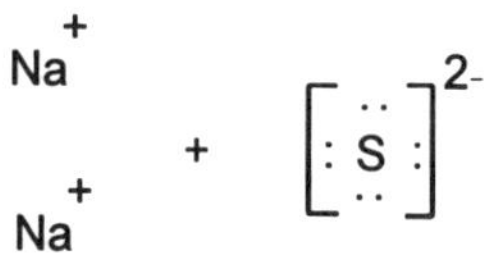

10.49 a. 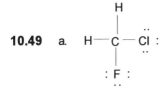 b.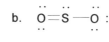

c. :O—N(=O)—O—H (with lone pairs shown) d. [:O—S(=O)(—O:)—O:]²⁻

10.51 a. In SO_2, 3 electron groups surround the central atom so the arrangement is trigonal planar.

b. In $SeCl_2$, 4 electron groups surround the central atom so the arrangement is tetrahedral.

c. In ClO_4^-, 4 electron groups surround the central atom so the arrangement is tetrahedral.

d. In PH_3, 4 electron groups surround the central atom so the arrangement is tetrahedral.

10.53 a. tetrahedral shape b. bent or angular shape
c. trigonal planar shape d. tetrahedral shape

10.55 H–Cl electronegativity difference = 3.0 - 2.1 = 0.9
P–Cl electronegativity difference = 3.0 - 2.1 = 0.9
N–Cl electronegativity difference = 3.0 - 3.0 = 0.0

The smallest difference in electronegativity, and thus the least polar bond, is N–Cl.

10.57 a. ionic (metal + nonmetal) b. covalent (two nonmetals)
c. ionic (metal + nonmetal) d. covalent (two nonmetals)

10.59 a. HF b. NF_3 c. H_2O

Solutions to Practice in Problem Analysis

10.1 The first step in determining the electron-dot formula of H_2S is to determine which atom is the central atom. Since an H atom cannot bond to more than one atom, it cannot be the central atom. Therefore, sulfur is the central atom. Next, calculate the total number of valence electrons for the molecule. Add up the valence electrons for each of the atoms in H_2S. The number of electrons contributed by each atom is equal to the group number of the element. Hydrogen is in Group IA with one valence electron, and sulfur is in Group VIA with six valence electrons (the total is 1 + 1 + 6 = 8). Connect the atoms together by single bonds, using either a pair of dots or a dash. Distribute electron dots to the atoms surrounding the central atom to satisfy the octet rule for them, and calculate the number of remaining valence electrons. Distribute these electrons to the central atom. If there are fewer than eight electrons around the central atom, convert one or more of the bonds to the central atom to multiple bonds by moving a lone pair on a surrounding atom to the bonding region between the atoms. Continue until the octet rule is obeyed for each atom in the molecule.

10.2 The first step in determining the electron-dot formula of $COCl_2$ is to determine which atom is the central atom. The most electropositive element is carbon, so it is the central atom. Next, calculate the total number of valence electrons for the molecule. Add up the valence electrons for each of the atoms in $COCl_2$ The number of electrons contributed by each atom is equal to the group number of the element. Carbon is in Group IVA with four valence electron, oxygen is in Group VIA with six valence electrons, and chlorine is in Group VIIA with seven valence electrons (the total is 4 + 6 + 7 + 7 = 24). Connect the atoms together by single bonds, using either a pair of dots or a dash. Distribute electron dots to the atoms surrounding the central atom to satisfy the octet rule for them, and calculate the number of remaining valence electrons. Distribute these electrons to the central atom. If there are fewer than eight electrons around the central atom, convert one or more of the bonds to the central atom to multiple bonds by moving a lone pair on a surrounding atom to the bonding region between the atoms. Continue until the octet rule is obeyed for each atom in the molecule.

Answers to the Practice Exam

1. a   2. d   3. e   4. e   5. b   6. c   7. e   8. d   9. a   10. c
11. d   12. a   13. e   14. b   15. b

11. THE GASEOUS STATE

■ Solutions to Exercises

Note on significant figures: When the final answer is written for the first time, it is written with one nonsignificant digit, and the least significant digit is underlined. The final answer is then rounded to the correct number of significant figures.

11.1 $0.712 \text{ atm} \times \dfrac{760 \text{ mmHg}}{1 \text{ atm}} = 54\underline{1}.1 = 541 \text{ mmHg}$

11.2 The quantities you need are

$$P_1 = 1.00 \text{ atm} \qquad P_2 = 2.10 \text{ atm}$$
$$V_1 = 20.0 \text{ L} \qquad V_2 = ?$$

Notice that the temperature, 23°C, is constant. Since the pressure increases, the volume must decrease. Insert these quantities into the equation

$$V_2 = V_1 \times \dfrac{P_1}{P_2}$$

This gives

$$V_2 = 20.0 \text{ L} \times \dfrac{1.00 \text{ atm}}{2.10 \text{ atm}} = 9.5\underline{2}4 = 9.52 \text{ L}$$

11.3 The quantities you need are

P_1 = 745 mmHg
V_1 = 1.32 L

→

P_2 = ?
V_2 = 1.89 L

Notice that the temperature, 18°C, is constant. Since the volume increases, the pressure must decrease. Insert these quantities into the equation

$$P_2 = P_1 \times \frac{V_1}{V_2}$$

This gives

$$P_2 = 745 \text{ mmHg} \times \frac{1.32 \text{ L}}{1.89 \text{ L}} = 52\underline{0}.3 = 5.20 \times 10^2 \text{ mmHg}$$

11.4 Convert the temperatures to the Kelvin scale:

initial temperature (T_1) = 19°C + 273 = 292 K
final temperature (T_2) = 25°C + 273 = 298 K

The quantities you need for this problem are

V_1 = 438 mL
T_1 = 292 K

→

V_2 = ?
T_2 = 298 K

Notice that there is no change in pressure, 742 mmHg, so it is constant. The equation you want is

$$V_2 = V_1 \times \frac{T_2}{T_1}$$

Since the temperature increases, the volume must also increase. This gives

$$V_2 = 438 \text{ mL} \times \frac{298 \text{ K}}{292 \text{ K}} = 44\underline{7}.0 = 447 \text{ mL}$$

11.5 First, convert the initial temperature to the Kelvin scale

initial temperature (T_1) = 50.0°C + 273 = 323 K

Next, identify the quantities you need:

V_1 = 10.0 mL V_2 = 20.0 mL
T_1 = 323 K T_2 = ?

Notice that the pressure, 736 mmHg, is not changing so is constant. The equation you want is

$$T_2 = T_1 \times \frac{V_2}{V_1}$$

Since the volume increases, the temperature must also increase. This gives

$$T_2 = 323 \text{ K} \times \frac{20.0 \text{ mL}}{10.0 \text{ mL}} = 646.0 \text{ K}$$

Finally, convert the final temperature to the Celsius scale.

°C = K - 273 = 646.0 K - 273 = 37$\underline{3}$.0 = 373°C

11.6 Calculate the mass of 22.4 L from

$$\frac{5.9 \text{ g}}{3.0 \text{ L}} \times \frac{22.4 \text{ L}}{1 \text{ mol}} = 4\underline{4}.0 = 44 \text{ g/mol}$$

Thus, the molar mass of carbon dioxide is 44 g/mol.

11.7 You must find the volume of chlorine at STP. First, convert the temperatures to kelvins.

initial temperature (T_1) = 255°C + 273 = 528 K

final temperature (T_2) = 0°C + 273 = 273 K

The equation you need is

$$V_2 = V_1 \times \frac{P_1}{P_2} \times \frac{T_2}{T_1}$$

The quantities you need for this problem are

P_1 = 5.14 atm	P_2 = 1.00 atm
V_1 = 119 mL	V_2 = ?
T_1 = 528 K	T_2 = 273 K

This gives

$$V_2 = 119 \text{ mL} \times \frac{5.14 \text{ atm}}{1.00 \text{ atm}} \times \frac{273 \text{ K}}{528 \text{ K}} = 316.3 \text{ mL}$$

The volume must be expressed in liters. This gives

$$316.3 \text{ mL} \times \frac{1 \text{ L}}{1000 \text{ mL}} = 0.3163 \text{ L}$$

The molar mass of chlorine can now be calculated using the molar volume. Since 1.00 g of chlorine is present, you get

$$\frac{1.00 \text{ g}}{0.3163 \text{ L}} \times \frac{22.4 \text{ L}}{1 \text{ mol}} = 70.\underline{8}1 = 70.8 \text{ g/mol}$$

11.8 First, change the temperature to the Kelvin scale.

T = 17°C + 273 = 290 K

The quantities that you need to use in the calculation are

P = ?	n = 0.500 moles
V = 3.06 L	R = 0.08206 L atm / (K mol)
	T = 290 K

The pressure can now be calculated.

$$P = \frac{nRT}{V} = \frac{0.500 \text{ mol} \times 0.08206 \text{ L atm}/(\text{K mol}) \times 290 \text{ K}}{3.06 \text{ L}}$$

$$= 3.8\underline{8}8 = 3.89 \text{ atm}$$

11.9 The quantities used in the calculation are

P = 0.886 atm	n = ?
V = 0.452 L	R = 0.08206 L atm / (K mol)
	T = 298 K

The moles of gas can now be calculated.

$$n = \frac{PV}{RT} = \frac{0.886 \text{ atm} \times 0.452 \text{ L}}{0.08206 \text{ L atm}/(\text{K mol}) \times 298 \text{ K}} = 0.016\underline{3}8 = 0.0164 \text{ mol}$$

The molar mass of CO_2 is 44.01 g/mol. Thus, the mass of gas is

$$0.01638 \text{ mol} \times \frac{44.01 \text{ g}}{1 \text{ mol}} = 0.72\underline{0}9 = 0.721 \text{ g}$$

11.10 First, convert the temperature to kelvins.

T = 25°C + 273 = 298 K

Next, calculate the pressure of each component of the mixture in the flask. The quantities that you need for carbon dioxide are

P = ?	n = 0.0200 moles
V = 2.00 L	R = 0.08206 L atm/(K mol)
	T = 298 K

Substitution leads to

$$P = \frac{nRT}{V} = \frac{0.0200 \text{ mol} \times 0.08206 \text{ L atm}/(K \text{ mol}) \times 298 \text{ K}}{2.00 \text{ L}}$$

$$= 0.244\underline{5} = 0.245 \text{ atm}$$

The quantities that you need for oxygen are

P = ?	n = 0.0400 moles
V = 2.00 L	R = 0.08206 L atm/(K mol)
	T = 298 K

Substitution leads to

$$P = \frac{nRT}{V} = \frac{0.0400 \text{ mol} \times 0.08206 \text{ L atm}/(K \text{ mol}) \times 298 \text{ K}}{2.00 \text{ L}}$$

$$= 0.489\underline{1} = 0.489 \text{ atm}$$

The total pressure in the flask is

$$P_T = P_{\text{carbon dioxide}} + P_{\text{oxygen}} = 0.2445 \text{ atm} + 0.4891 \text{ atm} = 0.733\underline{6} = 0.734 \text{ atm}$$

11.11 The molar mass of $CaCO_3$ is 100.09 g/mol. According to the balanced chemical equation, each mole of calcium carbonate used in the reaction generates 1 mol of carbon dioxide. So the amount of carbon dioxide in moles is

$$\text{mol } CO_2 = 15.0 \text{ g } CaCO_3 \times \frac{1 \text{ mol } CaCO_3}{100.09 \text{ g } CaCO_3} \times \frac{1 \text{ mol } CO_2}{1 \text{ mol } CaCO_3}$$

$$= 0.1499 \text{ mol } CO_2$$

The temperature of this gas in kelvins is 27°C + 273 = 300 K. The pressure must be converted to atmospheres.

$$P = 745 \text{ mmHg} \times \frac{1 \text{ atm}}{760 \text{ mmHg}} = 0.9803 \text{ atm}$$

The quantities to use in the equation are

P = 0.9083 atm	n = 0.1499 mol
V = ?	R = 0.08206 L atm/ (K mol)
	T = 300 K

Substitution leads to

$$V = \frac{nRT}{P} = \frac{0.1499 \text{ mol} \times 0.08206 \text{ L atm}/(K \text{ mol}) \times 300 \text{ K}}{0.9803 \text{ atm}}$$

$$= 3.7\underline{6}4 = 3.76 \text{ L}$$

■ Answers to Questions to Test Your Reading

11.1 The three states of matter are gases, liquids, and solids.

11.2 The four properties of all gases are

　i. Gases fill a container completely and uniformly.
　ii. Gases are compressible.
　iii. Gases have low densities.
　iv. Gases exert a uniform pressure on all inner surfaces of a container because it completely fills the container.

11.3 Pressure is force exerted on a unit area. Atmospheric pressure is the pressure exerted by the atmosphere, and is measured with a barometer. The height of the column of mercury in the barometer, in mm, is equal to the atmospheric pressure, in mmHg.

11.4 The atmosphere around the earth contains about 5×10^{18} kg of gaseous substances, which exerts a force on every part of the earth's surface because of the earth's gravitational attraction. This force per unit area is atmospheric pressure.

11.5 The average atmospheric pressure at sea level is 760 mmHg.

11.6 A manometer is a device that measures a gas's pressure in a container. The manometer is an open ended U-tube containing mercury (see Figure 11.5). The difference in the heights of mercury in the two arms of the U-tube indicates the pressure difference, in mmHg, between the gas in the container and the atmospheric pressure.

11.7 The pressure of a gas inside a perfectly elastic balloon is equal to the atmospheric pressure outside. Thus, the pressure inside the balloon would be 1 atm.

11.8 According to the kinetic molecular theory of gases, gases exert a pressure on any surface that they touch because the gas particles collide with the surface and push against it.

11.9 Boyle's law states that the volume of a fixed molar amount of gas at a given temperature is inversely proportional to the applied pressure.

11.10 Since, according to Boyle's law, at constant temperature, pressure and volume are inversely related. Thus, the lower the pressure, the larger the volume. Since the atmospheric pressure is lower at higher altitudes, the volume would be larger in Denver, Colorado.

11.11 The fourth postulate of the theory states that the average kinetic energy of gas particles is directly proportional to the Kelvin temperature. As a result, the average energy of a collision when one of the gas particles hits a wall of the container will not change when the volume decreases at constant temperature. However, the pressure will increase when the volume decreases because of more collisions with the container per unit area. Thus, pressure and volume should be inversely related as Boyle's law states.

11.12 Charles' law states that the volume of a fixed amount of gas is directly proportional to its temperature as long as the pressure is kept constant.

11.13 According to Charles' law, at constant pressure, the volume and temperature are directly related. Thus, the higher the temperature, the larger the volume. Therefore, the balloon would be larger on a hot summer day.

11.14 Since the pressure of a gas inside a non-rigid container must always equal the atmospheric pressure, the pressure of the gas will not change when the temperature changes. For a decrease in temperature, the average kinetic energy of the gas particles will decrease. Less powerful collisions with the walls of the container will occur and it will contract. Thus, when the temperature decreases, the volume will decrease. In accord with Charles' law, volume and temperature are directly proportional.

11.15 STP, or standard temperature and pressure, is an arbitrarily chosen reference state of exactly 0°C and 1 atmosphere pressure.

11.16 Avogadro's law can be stated in two different ways:

i. equal volumes of any two gases at the same temperature and pressure contain equal moles of molecules, and
ii. the volume of a gas is directly proportional to its amount in moles as long as the temperature and pressure are constant.

11.17 Since the dimensions of gas particles are very small compared to the very large average distance between the particles, the sizes of the gas particles is not a factor. Gases are mostly empty space.

11.18 The ideal gas law can be written as PV = nRT, where R is the molar gas constant.

11.19 Dalton's law of partial pressures states that the total pressure exerted by a gaseous mixture is the sum of the partial pressures of the components of the mixture. The partial pressure of a gas in a mixture is the pressure the gas would exert if it were in the container alone.

11.20 The kinetic molecular theory explains Dalton's law of partial pressures as follows. Because the attractions (or repulsions) between gas particles are negligible, the components of a gaseous mixture act independently, causing partial pressures that are proportional to their molar quantities.

■ Solutions to Practice Problems

Note on significant figures: When the final answer is written for the first time, it is written with one nonsignificant digit, and the least significant digit is underlined. The final answer is then rounded to the correct number of significant figures.

11.21 $1128 \text{ mmHg} \times \dfrac{1 \text{ atm}}{760 \text{ mmHg}} = 1.48\underline{4}2 = 1.484 \text{ atm}$

11.23 $V_2 = V_1 \times \dfrac{P_1}{P_2} = 2.60 \text{ L} \times \dfrac{768 \text{ mmHg}}{614 \text{ mmHg}} = 3.2\underline{5}2 = 3.25 \text{ L}$

11.25 $V_2 = V_1 \times \dfrac{P_1}{P_2} = 849 \text{ L} \times \dfrac{1.00 \text{ atm}}{21.6 \text{ atm}} = 39.\underline{3}1 = 39.3 \text{ L}$

11.27 $P_2 = P_1 \times \dfrac{V_1}{V_2} = 763 \text{ mmHg} \times \dfrac{845 \text{ mL}}{643 \text{ mL}} = 10\underline{0}3 = 1.00 \times 10^3 \text{ mmHg}$

11.29 $P_2 = P_1 \times \dfrac{V_1}{V_2} = 1.00 \text{ atm} \times \dfrac{145 \text{ L}}{165 \text{ L}} = 0.87\underline{8}8 = 0.879 \text{ atm}$

11.31 initial temperature $(T_1) = 25°C + 273 = 298 \text{ K}$
final temperature $(T_2) = 0°C + 273 = 273 \text{ K}$

$V_2 = V_1 \times \dfrac{T_2}{T_1} = 1.33 \times 10^3 \text{ L} \times \dfrac{273 \text{ K}}{298 \text{ K}} = 12\underline{1}8 = 1.22 \times 10^3 \text{ L}$

11.33 initial temperature $(T_1) = 26°C + 273 = 299 \text{ K}$
final temperature $(T_2) = 14°C + 273 = 287 \text{ K}$

$V_2 = V_1 \times \dfrac{T_2}{T_1} = 125 \text{ mL} \times \dfrac{287 \text{ K}}{299 \text{ K}} = 12\underline{0}.0 = 1.20 \times 10^2 \text{ mL}$

11.35 initial temperature $(T_1) = 0°C + 273 = 273 \text{ K}$

$T_2 = T_1 \times \dfrac{V_2}{V_1} = 273 \text{ K} \times \dfrac{3.22 \text{ L}}{2.22 \text{ L}} = 396.0 \text{ K}$

final temperature $(T_2) = 396.0 \text{ K} - 273 = 12\underline{3}.0 = 123°C$

11.37 initial temperature (T_1) = 30°C + 273 = 303 K

$$T_2 = T_1 \times \frac{V_2}{V_1} = 303 \text{ K} \times \frac{4.00 \text{ L}}{2.00 \text{ L}} = 606.0 \text{ K}$$

final temperature (T_2) = 606.0 K - 273 = 33$\underline{3}$.0 = 333°C

11.39 $\dfrac{15.3 \text{ g}}{12.2 \text{ L}} \times \dfrac{22.4 \text{ L}}{1 \text{ mol}} = 28.\underline{0}9 = 28.1$ g/mol

11.41 First, calculate the moles of argon. The temperature is 25°C + 273 = 298 K.

$$n = \frac{PV}{RT} = \frac{2.00 \text{ atm} \times 0.561 \text{ L}}{0.08206 \text{ L atm}/(\text{K mol}) \times 298 \text{ K}} = 0.04588 \text{ mol}$$

The molar mass is therefore

$$\frac{1.83 \text{ g}}{0.04588 \text{ mol}} = 39.\underline{8}9 = 39.9 \text{ g/mol}$$

11.43 The temperature is T = 23°C + 273 = 296 K. The pressure is

$$P = \frac{nRT}{V} = \frac{3.03 \text{ mol} \times 0.08206 \text{ L atm}/(\text{K mol}) \times 296 \text{ K}}{9.65 \text{ L}}$$

$$= 7.6\underline{2}7 = 7.63 \text{ atm}$$

11.45 The temperature is T = 100°C + 273 = 373 K. Plugging into the ideal gas equation gives

$$P = \frac{nRT}{V} = \frac{0.200 \text{ mol} \times 0.08206 \text{ L atm}/(\text{K mol}) \times 373 \text{ K}}{5.00 \text{ L}}$$

$$= 1.2\underline{2}4 = 1.22 \text{ atm}$$

124 ■ CHAPTER 11

11.47 The temperature is T = 55°C + 273 = 328 K. Plugging into the ideal gas equation gives

$$V = \frac{nRT}{P} = \frac{2.50 \text{ mol} \times 0.08206 \text{ L atm}/(K \text{ mol}) \times 328 \text{ K}}{1.50 \text{ atm}}$$

$$= 44.\underline{8}6 = 44.9 \text{ L}$$

11.49 The temperature is T = 75°C + 273 = 348 K. The pressure is given by

$$P = 775 \text{ mmHg} \times \frac{1 \text{ atm}}{760 \text{ mmHg}} = 1.020 \text{ atm}$$

The volume is given by

$$V = \frac{nRT}{P} = \frac{0.125 \text{ mol} \times 0.08206 \text{ L atm}/(K \text{ mol}) \times 348 \text{ K}}{1.020 \text{ atm}}$$

$$= 3.5\underline{0}0 = 3.50 \text{ L}$$

11.51 The temperature is T = 26°C + 273 = 299 K. The volume, in liters, is

$$125 \text{ mL} \times \frac{1 \text{ L}}{1000 \text{ mL}} = 0.125 \text{ L}$$

The pressure, in atmospheres, is given by

$$P = 745 \text{ mmHg} \times \frac{1 \text{ atm}}{760 \text{ mmHg}} = 0.9803 \text{ atm}$$

The moles of gas are

$$n = \frac{PV}{RT} = \frac{1.56 \text{ atm} \times 1.74 \text{ L}}{0.08206 \text{ L atm}/(K \text{ mol}) \times 298 \text{ K}}$$

$$= 0.00499\underline{4} = 4.99 \times 10^{-3} \text{ mol}$$

11.53 The temperature is given by T = 20°C + 273 = 293 K. The volume, in liters, is

$$V = 545 \text{ mL} \times \frac{1 \text{ L}}{1000 \text{ mL}} = 0.545 \text{ L}$$

The pressure, in atmospheres is

$$P = 760 \text{ mmHg} \times \frac{1 \text{ atm}}{760 \text{ mmHg}} = 1.00 \text{ atm}$$

The moles of gas are

$$n = \frac{PV}{RT} = \frac{1.00 \text{ atm} \times 0.545 \text{ L}}{0.08206 \text{ L atm}/(\text{K mol}) \times 293 \text{ K}} = 0.022\underline{6}7 = 0.0227 \text{ mol}$$

11.55 $$T = \frac{PV}{nR} = \frac{3.50 \text{ atm} \times 4.00 \text{ L}}{0.410 \text{ mol} \times 0.08206 \text{ L atm}/(\text{K mol})} = 416.1 \text{ K}$$

T = 416.1 K − 273 = 14\underline{3}.1 = 143°C

11.57 The moles of nitrogen, N_2 (molar mass 28.02 g/mol), is given by

$$28.0 \text{ g} \times \frac{1 \text{ mol}}{28.02 \text{ g}} = 0.9993 \text{ mol}$$

The pressure, in atmospheres, is given by

$$p = 755 \text{ mmHg} \times \frac{1 \text{ atm}}{760 \text{ mmHg}} = 0.9934$$

The temperature can now be determined.

$$T = \frac{PV}{nR} = \frac{0.9934 \text{ atm} \times 10.0 \text{ L}}{0.9993 \text{ mol} \times 0.08206 \text{ L atm}/(\text{K mol})} = 121.1 \text{ K}$$

T = 121.1 K − 273 = −15\underline{1}.9 = −152°C

11.59 The temperature is T = 15°C + 273 = 288 K. The partial pressure of oxygen is

$$P_{oxygen} = \frac{nRT}{V} = \frac{0.00125 \text{ mol} \times 0.08206 \text{ L atm}/(\text{K mol}) \times 288 \text{ K}}{2.00 \text{ L}}$$

$$= 0.01477 = 0.0148 \text{ atm}$$

The partial pressure of helium is

$$P_{helium} = \frac{nRT}{V} = \frac{0.00325 \text{ mol} \times 0.08206 \text{ L atm}/(\text{K mol}) \times 288 \text{ K}}{2.00 \text{ L}}$$

$$= 0.03841 = 0.0384 \text{ atm}$$

The total pressure is

$$P_{Tot} = P_{oxygen} + P_{helium} = 0.01477 \text{ atm} + 0.03840 \text{ atm} = 0.05317 = 0.0532 \text{ atm}$$

11.61 The first step is to calculate the number of moles of hydrogen peroxide. The molar mass of H_2O_2 is 34.02 g/mol.

$$\text{moles } H_2O_2 = 1.00 \text{ g} \times \frac{1 \text{ mol}}{34.02 \text{ g}} = 0.02939 \text{ mol } H_2O_2$$

The second step is to calculate the number of moles of oxygen by recognizing that two moles of hydrogen peroxide produce one mole of oxygen.

$$\text{moles } O_2 = 0.02939 \text{ mol } H_2O_2 \times \frac{1 \text{ mol}}{2 \text{ mol } H_2O_2} = 0.01470 \text{ mol } O_2$$

The final step is to use the ideal gas equation to calculate the volume. The temperature is T = 28°C + 273 = 301 K.

$$V = \frac{nRT}{P} = \frac{0.01470 \text{ mol} \times 0.08206 \text{ L atm}/(\text{K mol}) \times 301 \text{ K}}{0.867 \text{ atm}}$$

$$= 0.4188 = 0.419 \text{ L}$$

11.63 The first step is to calculate the number of moles of methane gas. The molar mass of CH_4 is 16.04 g/mol.

$$\text{moles } CH_4 = 10.0 \text{ g} \times \frac{1 \text{ mol}}{16.04 \text{ g}} = 0.6234 \text{ mol } CH_4$$

The second step is to calculate the number of moles of oxygen by recognizing that each mole of carbon dioxide requires two moles of oxygen.

$$\text{moles } O_2 = 0.6234 \text{ mol } CH_4 \times \frac{2 \text{ mol } O_2}{1 \text{ mol } CH_4} = 1.247 \text{ mol } O_2$$

The final step is to use the ideal gas equation to calculate the volume. The temperature is T = 23°C + 273 = 296 K.

$$V = \frac{nRT}{P} = \frac{1.247 \text{ mol} \times 0.08206 \text{ L atm}/(K \text{ mol}) \times 296 \text{ K}}{1.12 \text{ atm}}$$

$$= 27.0\underline{4} = 27.0 \text{ L}$$

11.65 The first step is to calculate the moles of oxygen using the ideal gas equation. The temperature is T = -12°C + 273 = 261 K. The pressure, in atmospheres, is

$$P = 664 \text{ mmHg} \times \frac{1 \text{ atm}}{760 \text{ mmHg}} = 0.8737 \text{ atm}$$

The moles of oxygen is given by

$$n = \frac{PV}{RT} = \frac{0.8737 \text{ atm} \times 10.8 \text{ L}}{0.08206 \text{ L atm}/(K \text{ mol}) \times 261 \text{ K}} = 0.4405 \text{ mol}$$

The next step is to determine the moles of magnesium by recognizing that one mole of oxygen reacts with two moles of magnesium. Finally, multiply by the molar mass of magnesium, 24.31 g/mol. Combining the last two steps gives

$$0.4405 \text{ mol } O_2 \times \frac{2 \text{ mol Mg}}{1 \text{ mol } O_2} \times \frac{24.31 \text{ g Mg}}{1 \text{ mol Mg}} = 21.\underline{4}2 = 21.4 \text{ g Mg}$$

11.67 The first step is to calculate the moles of sulfur dioxide, SO_2, at STP. This is given by

$$102 \text{ mL} \times \frac{1 \text{ L}}{1000 \text{ mL}} \times \frac{1 \text{ mol}}{22.4 \text{ L}} = 0.004554 \text{ mol } SO_2$$

The next step is to determine the moles of calcium oxide by recognizing that one mole of sulfur dioxide reacts with one mole of calcium oxide. Finally, multiply by the molar mass of calcium oxide, 56.08 g/mol. Combining the last two steps gives

$$0.004554 \text{ mol } SO_2 \times \frac{1 \text{ mol CaO}}{1 \text{ mol } SO_2} \times \frac{56.08 \text{ g CaO}}{1 \text{ mol CaO}}$$

$$= 0.255\underline{4} = 0.255 \text{ g CaO}$$

11.69 The balanced chemical equation is $BaCO_3(s) \xrightarrow{\Delta} BaO(s) + CO_2(g)$. The first step in the calculation is to determine the moles of carbon dioxide using the ideal gas equation. The temperature is T = 802°C + 273 = 1075 K.

$$n = \frac{PV}{RT} = \frac{0.955 \text{ atm} \times 19.4 \text{ L}}{0.08206 \text{ L atm}/(K \text{ mol}) \times 1075 \text{ K}} = 0.2100 \text{ mol}$$

The next step is to determine the moles of barium carbonate by recognizing that one mole of carbon dioxide is produced from one mole of barium carbonate. Finally, multiply by the molar mass of barium carbonate, 197.3 g/mol. Combining the last two steps gives

$$0.2100 \text{ mol } CO_2 \times \frac{1 \text{ mol } BaCO_3}{1 \text{ mol } CO_2} \times \frac{197.3 \text{ g } BaCO_3}{1 \text{ mol } BaCO_3}$$

$$= 41.4\underline{3} = 41.4 \text{ g } BaCO_3$$

Solutions to Additional Problems

11.71 The temperature is T = 23°C + 273 = 296 K. The volume is 12.5 mL, or 0.0125 L. The pressure, in atmospheres, is

$$P = 765 \text{ mmHg} \times \frac{1 \text{ atm}}{760 \text{ mmHg}} = 1.007 \text{ atm}$$

The moles of helium produced are

$$n = \frac{PV}{RT} = \frac{1.007 \text{ atm} \times 0.0125 \text{ L}}{0.08206 \text{ L atm}/(\text{K mol}) \times 296 \text{ K}} = 5.182 \times 10^{-4} \text{ mol He}$$

Next, calculate the number of helium atoms.

$$\text{atoms He} = 5.182 \times 10^{-4} \text{ mol He} \times \frac{6.022 \times 10^{23} \text{ atoms}}{1 \text{ mol He}}$$

$$= 3.121 \times 10^{20} \text{ atoms He}$$

Finally, you can calculate the number of atoms of the metal by recognizing that each metal atom results in one helium atom.

$$\text{metal atoms} = 3.121 \times 10^{20} \text{ atoms He} \times \frac{1 \text{ metal atom}}{1 \text{ atom He}}$$

$$= 3.12\underline{1} \times 10^{20} = 3.12 \times 10^{20} \text{ metal atoms}$$

11.73 You must first determine the partial pressure of the helium.

$$P_{\text{Helium}} = P_{\text{Total}} - P_{\text{oxygen}} = 3.00 \text{ atm} - 0.200 \text{ atm} = 2.800 \text{ atm}$$

Next, use the ideal gas law to find the number of moles. The temperature is T = 23°C + 273 = 296 K.

$$n = \frac{PV}{RT} = \frac{2.800 \text{ atm} \times 1.00 \text{ L}}{0.08206 \text{ L atm}/(\text{K mol}) \times 296 \text{ K}} = 0.115\underline{3} = 0.115 \text{ mol}$$

Solutions to Practice in Problem Analysis

11.1 This problem is uses Boyle's law to determine the volume change under constant temperature conditions. The initial pressure(P_1) is given as 743 mmHg and the final pressure(P_2) is 1158 mmHg. No initial volume is given. Therefore, using Boyle's law, you can only determine the ratio of the final volume (V_2) to the initial volume (V_1). Since the pressure is increasing, the volume will decrease.

11.2 This problem involves two separate reactions. The first is the combustion of hydrogen gas with oxygen to form water. The reaction is given by

$$2\,H_2\,(g) + O_2\,(g) \longrightarrow 2\,H_2O(l)$$

To determine the moles of water that are formed in the reaction, use the ideal gas equation to calculate the number of moles of hydrogen initially present. Then, the moles of water can be determined by recognizing that two moles of hydrogen react to form two moles of water.

The second reaction is the reaction of water with P_4O_{10} to form phosphoric acid. The reaction is given by

$$P_4O_{10}\,(s) + 6\,H_2O(l) \longrightarrow 4\,H_3PO_4\,(aq)$$

The mass of phosphoric acid formed can now be determined by multiplying the moles of water by the mole ratio of phosphoric acid to water (4 mol H_3PO_4 / 6 mol H_2O), and multiplying by the molar mass of phosphoric acid (97.99 g/mol).

Answers to the Practice Exam

1. d   2. c   3. d   4. c   5. a   6. d   7. c   8. b   9. c   10. a

11. a   12. d   13. d   14. a   15. d

12. LIQUIDS, SOLIDS, AND ATTRACTIONS BETWEEN MOLECULES

■ Solutions to Exercises

<u>Note on significant figures:</u> When the final answer is written for the first time, it is written with one nonsignificant digit, and the least significant digit is underlined. The final answer is then rounded to the correct number of significant figures.

12.1 Frost is the condensation of water vapor to ice without going through the liquid state. Thus, deposition has occurred.

12.2 First, determine the number of moles in 1.0 g of mercury (molar mass 200.59 g/mol).

$$1.0 \text{ g} \times \frac{1 \text{ mol}}{200.59 \text{ g}} = 0.00499 \text{ mol}$$

Next, calculate the heat that must be removed.

$$0.00499 \text{ mol} \times \frac{2.3 \text{ kJ}}{1 \text{ mol}} = 0.01\underline{1}4 = 0.011 \text{ kJ}$$

12.3 The number of moles in 1.0 g of mercury is 0.00499 mol (see Exercise 2). The heat required is

$$0.00499 \text{ mol} \times \frac{59.3 \text{ kJ}}{1 \text{ mol}} = 0.2\underline{9}6 = 0.30 \text{ kJ}$$

132 ■ CHAPTER 12

12.4 This is a three step process. The first step is to calculate the heat that must be removed to convert 5.0 g of gaseous mercury at 357°C to liquid mercury. The number of moles of mercury is (molar mass 200.59 g/mol)

$$5.0 \text{ g} \times \frac{1 \text{ mol}}{200.59 \text{ g}} = 0.0249 \text{ mol}$$

The heat removed in this step is (the molar heat of vaporization is 59.3 kJ/mol)

$$0.0249 \text{ mol} \times \frac{59.3 \text{ kJ}}{1 \text{ mol}} = 1.48 \text{ kJ}$$

The second step is to calculate the heat that must be removed to lower the temperature from 357°C to -39°C, a change in temperature of 396°C (the specific heat is 0.14 J/g°C).

$$5.0 \text{ g} \times \frac{0.14 \text{ J}}{1 \text{ g°C}} \times 396°C = 277 \text{ J} = 0.277 \text{ kJ}$$

The third step is to calculate the heat that must be removed to change the liquid mercury into solid mercury at -39°C (the molar heat of fusion is 2.3 kJ/mol).

$$0.0249 \text{ mol} \times \frac{2.3 \text{ kJ}}{1 \text{ mol}} = 0.0573 \text{ kJ}$$

The overall heat that must be removed is

$$1.48 \text{ kJ} + 0.277 \text{ kJ} + 0.0573 \text{ kJ} = 1.8\underline{1} = 1.8 \text{ kJ}$$

12.5 According to Table 12.4, the vapor pressure of water at 15°C is 12.8 mmHg. The pressure of the gas is obtained from the total pressure as follows.

$$P = P_{Total} - P_{Water\ Vapor} = 735 \text{ mmHg} - 12.8 \text{ mmHg} = 722.2 \text{ mmHg}$$

Convert this result to atmospheres, the volume to liters, and the temperature to kelvins.

$$T = 15°C + 273 = 288 \text{ K}.$$

(continued)

LIQUIDS, SOLIDS, AND ATTRACTIONS BETWEEN MOLECULES ■ 133

$$P = 722.2 \text{ mmHg} \times \frac{1 \text{ atm}}{760 \text{ mmHg}} = 0.9503 \text{ atm}$$

$$V = 203 \text{ mL} \times \frac{1 \text{ L}}{1000 \text{ mL}} = 0.203 \text{ L}$$

Solve for the moles of gas using the ideal gas equation.

$$n = \frac{PV}{RT} = \frac{0.9503 \text{ atm} \times 0.203 \text{ L}}{0.08206 \text{ L atm}/(\text{K mol}) \times 288 \text{ K}} = 0.008163 = 0.00816 \text{ mol}$$

12.6 Both hydrogen (H_2) and bromine (Br_2) are nonpolar molecules. Therefore, in the liquid state only dispersion forces are present. Hydrogen bromide (HBr) is polar. Thus, in addition to dispersion forces, there are also dipole-dipole forces present in the liquid state of HBr.

12.7 Dispersion forces increase with increasing molecular weight. The molecular weights of CH_4, C_2H_6, and C_3H_8 are 16 amu, 30 amu, and 44 amu, respectively. Each of the three molecules is nonpolar so there are no dipole-dipole forces present. There is also no hydrogen bonding in any of the molecules. Thus, the order of increasing attractive forces is the same as the molecular weight. We conclude that the vapor pressure is greater where the forces are less. Thus, in order of increasing vapor pressure, we can conclude $C_3H_8 < C_2H_6 < CH_4$.

12.8 Both methane (CH_4) and silane (SiH_4) are symmetrical and nonpolar. Thus, the only forces present are dispersion forces. Since the dispersion forces increase with increasing molecular weight, there are more forces in silane (molecular weight 32.12 amu) than in methane (molecular weight 16.04 amu). Also, vapor pressure decreases with increasing forces so the vapor pressure of silane is less than for methane. Thus, the boiling point of SiH_4 should be higher.

12.9 Zinc (Zn) exists as a metal; it is a metallic solid. Sodium bromide (NaBr) exists as an ionic solid. Methane (CH_4) is a molecular substance and thus it freezes as a molecular solid.

12.10 Magnesium chloride ($MgCl_2$) exists as an ionic solid. Methanol (CH_3OH) is a molecular substance and freezes as a molecular solid. Argon (Ar) is a noble gas and thus freezes as an atomic solid. Argon is a gas at room temperature and has a very low melting point. Methanol is a liquid at room temperature so its melting point is higher than for argon. Magnesium chloride is a solid at room temperature and, since it is an ionic compound, it has a high melting point. Thus, in order of increasing melting point, we can conclude Ar < CH_3OH < $MgCl_2$.

Answers to Questions to Test Your Reading

12.1 A gas has neither a fixed volume nor a fixed shape. A liquid has a fixed volume but it does not have a specific shape. A solid has a fixed volume and a fixed shape.

12.2 A change of state occurs when a substances changes from a solid, liquid, or gas into a different state. An example is when ice melts; water in the solid state has changed to water in the liquid state. This is called melting, or fusion.

12.3 The temperature is the principal factor that determines the state of matter, even though pressure also can be a factor. For example, at 1 atm pressure, water exists as a solid below 0°C, a liquid above 0°C and below 100°C, and a gas above 100°C. A change in temperature can trigger a change in state.

12.4 The different changes of state are as follows:

Melting (or fusion), which is the change of a solid to a liquid. The melting of ice is an example: $H_2O(s) \longrightarrow H_2O(l)$

Freezing, which is the change of a liquid to a solid. The freezing of water is an example: $H_2O(l) \longrightarrow H_2O(s)$

Vaporization, which is the change of a liquid into a gas. The vaporization of water is an example: $H_2O(l) \longrightarrow H_2O(g)$

Condensation, which is the change of a gas to a liquid, or a gas to a solid. An example is the condensation of water: $H_2O(g) \longrightarrow H_2O(l)$

Sublimation, which is the change of a solid to a gas. The sublimation of water is an example: $H_2O(s) \longrightarrow H_2O(g)$

12.5 At 25°C, water (H_2O) and mercury (Hg) are liquids. Sodium chloride (NaCl) and quartz (SiO_2) are solids at 25°C.

12.6 The melting point of ice is 0°C, and at this temperature, ice and water can coexist indefinitely in equilibrium, as long as no more heat is added. The normal boiling point of water (at 760 mmHg) is 100°C. Since 120°C is above the normal boiling point of water, liquid water and steam cannot coexist indefinitely at this temperature.

12.7 The constant temperature at which a solid can coexist with its liquid is the melting point of the solid, and the freezing point of the liquid. For a given substance these are always identical.

12.8 The normal melting point of a solid is the constant temperature, at 1 atm pressure, at which the solid changes to a liquid.

12.9 The normal boiling point of a liquid is the constant temperature, at 1 atm pressure, at which the liquid changes to a gas.

12.10 The normal freezing point of water is 0°C and the normal boiling point is 100°C.

12.11 The molar heat of fusion is the energy required to melt (fuse) one mole of a solid. The molar heat of vaporization is the energy required to vaporize one mole of a liquid.

12.12 18 g (1 mole) of steam at 100°C has as much heat as liquid water at 100°C, plus an additional 40.7 kJ (the heat of vaporization). After 40.7 kJ of heat has been removed from the steam, it condenses to water at 100°C. Thus, the steam can melt more ice.

12.13 When a liquid evaporates, it undergoes a phase change from liquid to gas. This requires that energy (heat) be absorbed from the remaining liquid. Thus, by removing heat from the liquid, it becomes cooler. As far as our bodies are concerned, water evaporates from our skin and removes heat in the process thereby cooling our bodies. This is called perspiration.

12.14 The vapor pressure of a liquid is the pressure exerted by the liquid's vapor at equilibrium. Vapor pressure depends on the temperature. As the temperature increases, the vapor pressure increases.

12.15 Water will boil at 50°C if the atmospheric pressure is equal to the vapor pressure of water at this temperature, which is 92.5 mmHg.

12.16 Below the boiling point, a liquid evaporates principally from its surface. In a closed container, a small quantity will evaporate, and then evaporation will appear to stop. This is because condensation is also occurring. The rate of evaporation and the rate of condensation will eventually become equal. This is called equilibrium.

12.17 Water droplets are spherical because a sphere has the smallest area per unit volume, which reduces the surface tension.

12.18 Dispersion forces are the weak attractive forces resulting from small, instantaneous dipoles that occur because of the varying positions of the electrons during their motion about nuclei. At any one instant in a molecule or atom, the electrons may not be uniformly distributed about the nuclei. For a brief instant, one side will have a small positive charge and the other side will have a small negative charge. This is called an instantaneous dipole. This instantaneous dipole can cause additional instantaneous dipoles in other nearby molecules. The result is that the two instantaneous dipoles attract each other as a dispersion force.

(continued)

Since dispersion forces occur in the liquid state of all molecules, both liquid nitrogen and liquid water will experience these forces. Because N_2 is nonpolar, dispersion forces are the only type of intermolecular force present, and N_2 is a gas at room temperature. However, H_2O is polar and other forces, including hydrogen bonding are present, so dispersion forces play a small role in the properties of water.

12.19 A dipole-dipole force is a permanent attractive intermolecular force resulting from the interaction of the positive end of one molecule with the negative end of another. These forces are important in any molecule that is unsymmetric, and therefore polar. An example of a liquid in which these forces are important is HCl.

12.20 Hydrogen bonding occurs between two polar molecules containing a hydrogen atom covalently bonded to an atom of either nitrogen, oxygen, or fluorine. An example of a liquid in which these forces are important is methanol, CH_3OH.

12.21 If water was a linear molecule, it would be symmetric, and thus nonpolar.

12.22 A crystalline solid is a solid that is composed of one or more crystals with each crystal having a well-defined, ordered arrangement of the structural units. Examples are sodium chloride and sucrose. An amorphous solid is a solid that lacks order, and as a result, it does not have a well-defined arrangement of the structural units. Glass is an example.

12.23 An atomic solid is a solid containing atoms from a nonmetallic element. An example is neon. A metallic solid is a solid containing atoms from a metallic element, such as copper. A covalent network solid is a solid containing atoms held in large networks or chains by covalent bonds. An example is carbon. A molecular solid is a solid consisting of molecules. Water (ice) is an example. An ionic solid is a solid consisting of ions. Sodium chloride is an example.

12.24 Ionic solids are held together by strong attractive forces that exist between oppositely charged ions. These forces are much stronger than those that exist between the partial positive and negative charges found in molecular solids. These strong forces cause ionic solids to have high melting points. Atomic solids come in a variety of types. Some are held together by only dispersion forces and have very low melting points. Noble gases are of this type. Metallic solids, which are also atomic solids, are held together by stronger forces and have relatively high melting points. A third type of atomic solid is for certain nonmetals like carbon and silicon. These are covalently bonded solids and have high melting points.

12.25 Covalent network solids have relatively high melting points because the strong covalent bonds must be broken.

12.26 Molecular solids can have either relatively low or relatively high melting points because, in addition to dispersion forces, some compounds include dipole-dipole forces or hydrogen bonding (high melting points), and some contain only dispersion forces (low melting points).

Solutions to Practice Problems

Note on significant figures: When the final answer is written for the first time, it is written with one nonsignificant digit, and the least significant digit is underlined. The final answer is then rounded to the correct number of significant figures.

12.27 a. Chlorine gas liquefies. This is called condensation.

b. Molten lava becomes a solid. This is freezing, or fusion.

12.29 First, calculate the number of moles in 75.0 g of ice. The molar mass of ice is 18.02 g/mol.

$$75.0 \text{ g} \times \frac{1 \text{ mol}}{18.02 \text{ g}} = 4.162 \text{ mol}$$

Next, calculate the heat required by using the molar heat of fusion.

$$4.162 \text{ mol} \times \frac{6.01 \text{ kJ}}{1 \text{ mol}} = 25.0\underline{1} = 25.0 \text{ kJ}$$

12.31 First, calculate the number of moles in 26.4 g of ammonia (NH_3). The molar mass of ammonia is 17.03 g/mol.

$$26.4 \text{ g} \times \frac{1 \text{ mol}}{17.03 \text{ g}} = 1.550 \text{ mol}$$

Next, calculate the heat required to melt the ammonia.

$$1.550 \text{ mol} \times \frac{5.65 \text{ kJ}}{1 \text{ mol}} = 8.7\underline{5}8 = 8.76 \text{ kJ}$$

12.33 First, calculate the number of moles in 75.0 g of water, molar mass 18.02 g/mol.

$$75.0 \text{ g} \times \frac{1 \text{ mol}}{18.02 \text{ g}} = 4.162 \text{ mol}$$

Next, calculate the heat required.

$$4.162 \text{ mol} \times \frac{40.7 \text{ kJ}}{1 \text{ mol}} = 16\underline{9}.4 = 169 \text{ kJ}$$

12.35 First, determine the moles of gold (molar mass 196.97 g/mol).

$$26.4 \text{ g} \times \frac{1 \text{ mol}}{196.97 \text{ g}} = 0.1340 \text{ mol}$$

Next, calculate the heat released.

$$0.1340 \text{ mol} \times \frac{310. \text{ kJ}}{1 \text{ mol}} = 41.\underline{5}4 = 41.5 \text{ kJ}$$

12.37 This is a three-step process. The first step is to calculate the heat required to change 12.2 g of ice to liquid water at 0°C. The moles of water are (molar mass 18.02 g)

$$12.2 \text{ g} \times \frac{1 \text{ mol}}{18.02 \text{ g}} = 0.6770 \text{ mol}$$

The heat required for this step is

$$0.6770 \text{ mol} \times \frac{6.01 \text{ kJ}}{1 \text{ mol}} = 4.07 \text{ kJ}$$

The second step is to calculate the heat required to change the water from 0°C to 100°C, a temperature change of 100°C.

$$12.2 \text{ g} \times \frac{4.184 \text{ J}}{1 \text{ g°C}} \times 100°C = 5104 \text{ J} = 5.10 \text{ kJ}$$

The third step is to calculate the heat required to change the water at 100°C to steam.

$$0.6770 \text{ mol} \times \frac{40.7 \text{ kJ}}{1 \text{ mol}} = 27.6 \text{ kJ}$$

The total heat required for the process is

$$4.07 \text{ kJ} + 5.10 \text{ kJ} + 27.6 \text{ kJ} = 36.\underline{7}7 = 36.8 \text{ kJ}$$

12.39 This is a two-step process. First, find the heat needed to change 33.3 g of water at 37°C to water at 100°C, a temperature change of 63°C.

$$33.3 \text{ g} \times \frac{4.184 \text{ J}}{1 \text{ g} \cdot °C} \times 63°C = 8780 \text{ J} = 8.78 \text{ kJ}$$

The second step is to calculate the heat needed to vaporize the water at 100°C. The moles of water are (molar mass 18.02 g)

$$33.3 \text{ g} \times \frac{1 \text{ mol}}{18.02 \text{ g}} = 1.848 \text{ mol}$$

The heat needed is

$$1.848 \text{ mol} \times \frac{40.7 \text{ kJ}}{1 \text{ mol}} = 75.21 \text{ kJ}$$

The total heat required for the process is

8.78 kJ + 75.21 kJ = 83.$\underline{9}$9 = 84.0 kJ

12.41 The partial pressure of hydrogen is obtained from the total pressure as follows

$$P_{Hydrogen} = P_{Total} - P_{Water\ Vapor} = 752 \text{ mmHg} - 31.8 \text{ mmHg}$$

$$= 720.2 = 720. \text{ mmHg}$$

The temperature, in kelvins, is T = 30°C + 273 = 303 K. The volume is 232 mL, or 0.232 L. The pressure, in atmospheres, is

$$P = 720.2 \text{ mmHg} \times \frac{1 \text{ atm}}{760 \text{ mmHg}} = 0.9476 \text{ atm}$$

Solve for the moles of hydrogen using the ideal gas equation.

$$n = \frac{PV}{RT} = \frac{0.9476 \text{ atm} \times 0.232 \text{ L}}{0.08206 \text{ L atm}/(K \text{ mol}) \times 303 \text{ K}}$$

$$= 0.008\underline{8}42 = 0.00884 \text{ mol}$$

12.43 a. The only attractive forces present in radon (Rn) are dispersion forces.

b. Hydrogen fluoride (HF) is a polar molecule. Since the hydrogen atom is bonded to a fluorine atom, there will be hydrogen bonding. Also, there will be dispersion forces.

c. Hydrogen iodide (HI) is polar and has dipole-dipole forces as well as dispersion forces.

12.45 Dispersion forces increase with increasing molecular weight. Since each of the molecules is a nonpolar tetrahedral molecule with only dispersion forces present, the order is $CCl_4 < SiCl_4 < GeCl_4$.

12.47 Both CH_4 and CCl_4 are nonpolar tetrahedral molecules so dispersion forces are the only attractive forces present. Since dispersion forces increase with increasing molar mass, the lower vapor pressure would be for CCl_4, where the dispersion forces are stronger.

12.49 Boiling point increases with increasing intermolecular forces. This is because the stronger the attractive forces the lower the vapor pressure and therefore the higher the temperature necessary to reach boiling. Since the three compounds are all nonpolar tetrahedral molecules and only have dispersion forces present, the dispersion forces increase with increasing molar mass. Thus, the order of increasing boiling point is $CCl_4 < SiCl_4 < GeCl_4$.

12.51 Hydrogen selenide (H_2Se) and water (H_2O) are both polar and have dipole-dipole forces. However, water also has hydrogen bonding and hydrogen selenide does not. Therefore, water has the stronger forces and the higher boiling point.

12.53 P_4 is a molecular substance. Therefore, it freezes as a molecular solid.

12.55 Barium chloride ($BaCl_2$) is an ionic solid.

12.57 Since both water (H_2O) and hydrogen selenide (H_2Se) are molecular substances, they would both freeze as molecular solids. The stronger attractive forces are in water due to hydrogen bonding. Thus, water has the higher boiling point and the higher melting point.

12.59 Nitrogen (N_2) is a gas at room temperature while phosphorus (P_4) is a solid. Therefore, phosphorus has the higher melting point.

LIQUIDS, SOLIDS, AND ATTRACTIONS BETWEEN MOLECULES ■ 141

■ Solutions to Additional Problems

12.61 We must first calculate the heat that is released when 525 g of water freezes at 0°C. The number of moles of water is

$$525 \text{ g} \times \frac{1 \text{ mol}}{18.02 \text{ g}} = 29.13 \text{ mol}$$

The heat released is

$$29.13 \text{ mol} \times \frac{6.01 \text{ kJ}}{1 \text{ mol}} = 175.1 \text{ kJ}$$

Since the molar heat of fusion for CCl_2F_2 is 17.4 kJ/mol, the moles of this compound required to absorb the heat is

$$175.1 \text{ kJ} \times \frac{1 \text{ mol}}{17.4 \text{ kJ}} = 10.06 \text{ mol}$$

The molar mass of CCl_2F_2 is 120.91 g/mol. Therefore, the mass required is

$$10.06 \text{ mol} \times \frac{120.91 \text{ g}}{1 \text{ mol}} = 12\underline{1}6 = 1.22 \times 10^3 \text{ g}$$

12.63 The heat of vaporization is a measure of the energy required to overcome the attractive forces in the substance. The stronger the forces, the higher the heat of vaporization. Since both Cl_2 and H_2 are nonpolar and only have dispersion forces present, the forces increase with increasing molecular mass. Thus Cl_2 should have the higher heat of vaporization, and it does.

Solutions to Practice in Problem Analysis

12.1 In order to solve this problem, it will be necessary to compare the heat absorbing capacities and states of matter for the two compounds. Chlorofluorocarbons are gases at normal temperatures, and are compressed into liquids as part of the refrigeration cycle. The liquid form of the chlorofluorocarbon absorbs heat and is converted back into a gas during the process. The amount of heat absorbed during the cycle is due to the warming of the liquid from the initial temperature up to its boiling point, the vaporization of the liquid to a gas at the boiling point, and the heating of the gas to the final temperature. The gas is then recondensed by a mechanical pump back into a liquid and the cycle is repeated.

Water is a liquid at room temperature. The liquid water would have to be at an initial temperature below room temperature to act as a coolant. The problem would be how to cool the water back down to its initial state to repeat the cycle. Unlike a gas, water cannot be compressed, so the heat cannot be removed by a mechanical pump. You would require a continuous supply of cold water, and a place to dump the heated water. Therefore, there would be no easy way to use water for refrigerator coolants. This process is used in some power plants. The plant is built on the bank of a river, lake, or ocean, and uses it as both a source of cold water and a place to dump the heated water.

12.2 The cooling properties of an alcohol rub are based of the evaporation of the alcohol from the skin. As the alcohol evaporates, it undergoes a phase change from liquid to gas. During the process, heat is absorbed from the body, lowering the temperature of the skin. As a result, the patient would experience a cooling effect from the evaporation of the alcohol.

Answers to the Practice Exam

1. a 2. c 3. e 4. e 5. a 6. a 7. a 8. e 9. e 10. e
11. a 12. b 13. b 14. a 15. c

13. SOLUTIONS

■ Solutions to Exercises

Note on significant figures: When the final answer is written for the first time, it is written with one nonsignificant digit, and the least significant digit is underlined. The final answer is then rounded to the correct number of significant figures.

13.1 Substitute into the defining equation for mass percent of solute:

$$\text{mass percent of solute} = \frac{\text{mass of solute}}{\text{mass of solution}} \times 100\%$$

$$= \frac{5.4 \text{ g}}{83.5 \text{ g}} \times 100\% = 6.\underline{4}7 = 6.5\%$$

13.2 You first calculate the mass of the solution:

mass of solution = mass of solute + mass of solvent = 5.4 g + 83.5 g = 88.9 g

Next, you calculate the mass percent of the solute:

$$\frac{5.4 \text{ g}}{88.9 \text{ g}} \times 100\% = 6.\underline{0}7 = 6.1\%$$

144 ■ CHAPTER 13

13.3 Start with the equation for mass of solute:

$$\text{mass of solute} = \frac{\text{mass percent of solute} \times \text{mass of solution}}{100\%}$$

Substitute into this equation to obtain the mass of solute:

$$\text{mass of solute} = \frac{5.0\% \times 42.5 \text{ g}}{100\%} = 2.\underline{1}3 = 2.1 \text{ g}$$

mass of water = 42.5 g - 2.13 g = 40.$\underline{3}$8 = 40.4 g

13.4 In order to obtain the moles of $NiSO_4$, you need the molecular weight, 154.75 amu. Thus, 1 mol of $NiSO_4$ = 154.75 g. Hence,

$$1.58 \text{ g} \times \frac{1 \text{ mol NiSO}_4}{154.75 \text{ g NiSO}_4} = 0.01021 \text{ mol NiSO}_4$$

The volume of the solution is 61.2 mL, or 0.0612 L. Thus,

$$\text{molarity} = \frac{0.01021 \text{ mol}}{0.0612 \text{ L}} = 0.16\underline{6}8 = 0.167 \text{ M}$$

13.5 First, convert the volume to moles using the molarity (250 mL = 0.250 L).

$$0.250 \text{ L AgNO}_3 \text{ solution} \times \frac{0.150 \text{ mol AgNO}_3}{1 \text{ L AgNO}_3 \text{ solution}} = 0.03750 \text{ mol AgNO}_3$$

Next, use the formula weight of $AgNO_3$ (169.88 amu) to find the mass.

$$0.03750 \text{ mol AgNO}_3 \times \frac{169.88 \text{ g AgNO}_3}{1 \text{ mol AgNO}_3} = 6.3\underline{7}1 = 6.37 \text{ g}$$

SOLUTIONS ■ 145

13.6 First, obtain the moles of NaOH. The volume is 48.1 mL or 0.0481 L.

$$0.0481 \text{ L NaOH solution} \times \frac{0.148 \text{ mol NaOH}}{1 \text{ L NaOH solution}} = 0.007119 \text{ mol NaOH}$$

The molecular weight of $HC_2H_3O_2$ is 60.03 amu. Using the balanced chemical equation gives:

$$0.007119 \text{ mol NaOH} \times \frac{1 \text{ mol } HC_2H_3O_2}{1 \text{ mol NaOH}} \times \frac{60.03 \text{ g } HC_2H_3O_2}{1 \text{ mol } HC_2H_3O_2}$$

$$= 0.427\underline{4} = 0.427 \text{ g}$$

13.7 First, obtain the equivalents of NaOH. The volume is 48.1 mL or 0.0481 L.

$$0.0481 \text{ L NaOH solution} \times \frac{0.148 \text{ eq NaOH}}{1 \text{ L NaOH solution}} = 0.007119 \text{ eq NaOH}$$

The equivalents of acetic acid initially in the flask is the same. Since 1 equivalent of $HC_2H_3O_2$ equals 1 mole of $HC_2H_3O_2$, the flask contains 0.007119 mol $HC_2H_3O_2$. Since the molecular weight of $HC_2H_3O_2$ is 60.03 amu, the mass of acetic acid in the vinegar sample is

$$0.007119 \text{ mol } HC_2H_3O_2 \times \frac{60.03 \text{ g } HC_2H_3O_2}{1 \text{ mol } HC_2H_3O_2} = 0.427\underline{4} = 0.427 \text{ g}$$

13.8 The amount of $AgNO_3$ in the dilute solution is

$$1.00 \text{ L AgNO}_3 \text{ solution} \times \frac{0.10 \text{ mol AgNO}_3}{1 \text{ L AgNO}_3 \text{ solution}} = 0.100 \text{ mol AgNO}_3$$

This is also the moles of $AgNO_3$ in the concentrated solution. Convert this to liters of solution using the molarity of the solution.

$$0.100 \text{ mol AgNO}_3 \times \frac{1 \text{ L solution}}{2.5 \text{ mol AgNO}_3} = 0.040\underline{0} = 0.040 \text{ L} = 40. \text{ mL}$$

CHAPTER 13

13.9 Substitute into the dilution equation to get

$$1.2 \text{ mol/L} \times V_i = 0.25 \text{ mol/L} \times 0.50 \text{ L}$$

Solving for V_i gives

$$V_i = \frac{0.25 \text{ mol/L} \times 0.50 \text{ L}}{1.2 \text{ mol/L}} = 0.1\underline{0}4 = 0.10 \text{ L} = 1.0 \times 10^2 \text{ mL}$$

Therefore, 1.0×10^2 mL of concentrated solution should be used to prepare the dilute solution.

13.10 First, find the moles of ethylene glycol, $C_2H_6O_2$, molecular weight 62.07 amu.

$$34.2 \text{ g } C_2H_6O_2 \times \frac{1 \text{ mol } C_2H_6O_2}{62.07 \text{ g } C_2H_6O_2} = 0.5510 \text{ mol } C_2H_6O_2$$

Using the molality formula with 250.0 g, or 0.2500 kg, of solvent gives

$$\text{molarity} = \frac{1 \text{ mol } C_2H_6O_2}{0.2500 \text{ kg solvent}} = 2.2\underline{0}4 = 2.20 \text{ m}$$

13.11 You must first find the molality of the solution. Sucrose, $C_{12}H_{22}O_{11}$, has a molecular weight 342.30 amu. Find the moles of sucrose

$$23.4 \text{ g } C_{12}H_{22}O_{11} \times \frac{1 \text{ mol } C_{12}H_{22}O_{11}}{342.30 \text{ g } C_{12}H_{22}O_{11}} = 0.06836 \text{ mol } C_{12}H_{22}O_{11}$$

Using the molality formula with 125 g, or 0.125 kg, solvent gives

$$\text{molarity} = \frac{0.06836 \text{ mol } C_{12}H_{22}O_{11}}{0.125 \text{ kg solvent}} = 0.5469 \text{ m}$$

You can now find the temperature change, ΔT_f.

$$\Delta T_f = K_f m = (1.86°C/m) \times (0.5469 \text{ m}) = 1.017°C$$

Therefore, the freezing point of the solution is $0.00°C - 1.017°C = -1.0\underline{1}7 = -1.02°C$

13.12 The boiling point of the solution in Exercise 13.11 (molality = 0.5469 m) is found by first finding ΔT_b:

$$\Delta T_b = K_b m = (0.52°C/m) \times (0.5469\ m) = 0.284°C$$

The boiling point is 100.00°C + 0.284°C = 100.2$\underline{8}$4 = 100.28°C

■ Answers to Questions to Test Your Reading

13.1 A solution is a homogeneous mixture of two or more substances. The solvent is the substance in a solution that dissolves another substance; if it is not clear which substance does the dissolving, it is the substance in the solution that is in greater amount. A solute is either a substance in a solution considered to have been dissolved by a solvent, or else a substance in a solution that is in smaller amount. In a sodium chloride and water mixture, the water is the solvent, sodium chloride is the solute, and the mixture is the solution.

13.2 Two liquids that mix completely in one another to form a solution, no matter what the proportions of liquids, are said to be miscible.

13.3 Two liquids that separate into two distinct layers after mixing are said to be immiscible. No matter how much you stir, they will not mix together.

13.4 a. Air is an example of a gaseous solution. It contains 78% N_2, 21% O_2, and also Ar, CO_2, and H_2O in trace amounts.

b. An example of a liquid solvent and gaseous solute is carbonation in beverages. This is carbon dioxide gas dissolved in a water solution.

c. A mixture of the appropriate amounts of solid sodium with solid potassium forms a liquid solution.

d. An alloy, such as jewelry gold, which is a mixture of gold and silver, is a solid solution.

e. Gasoline is an example of a nonaqueous solution.

13.5 a. A mixture of helium and oxygen is a gaseous solution. The solute and solvent are both gases.

b. Brass (copper and zinc) is a solid solution, an alloy. The solute and the solvent are both solids.

c. Antifreeze and water is a liquid solution. The solute and the solvent are both liquids.

d. Brine is a mixture of salt in water, an aqueous solution. The solute is a solid and the solvent is a liquid, water.

e. Chlorine (in the gaseous state) dissolves in water as a gaseous solute and a liquid solvent.

13.6 A compound is a pure substance of fixed composition by mass. An example is sodium chloride, which is 39.3% sodium and 60.7% chlorine by mass. A solution is two or more substances mixed in variable proportions. A compound and a solution are both homogeneous. However, a solution can be separated into its pure components by a physical process like distilling, while a compound can only be separated into elements by a chemical process.

13.7 In a bucket of sand and water, the sand will settle out because the particles are large enough to be affected by gravitational forces. Since sand is more dense than water it falls to the bottom of the bucket. Solute molecules dissolve to form individual molecules or ions that are stabilized by the solvent molecules so that gravity no longer affects them as a principal force. The individual particles are in constant, random motion due to collisions with other particles.

13.8 The solubility of lithium hydroxide in water at 20°C is 12.4 g per 100 mL of water. This means that a 100 mL volume of water dissolves a maximum of 12.4 g of lithium hydroxide at 20°C. Any more lithium hydroxide that you add would settle to the bottom.

13.9 The solution you obtain after you have dissolved the maximum amount of solute in a solvent at a given temperature is a saturated solution. If you take a small quantity of the solute and add it to the saturated solution, it will simply fall to the bottom of the vessel. In an unsaturated solution, it would dissolve.

13.10 The solubility of lithium hydroxide in water at 20°C is 12.4 g. Since 11.0 g is less than the maximum amount per 100 mL of water, it is an unsaturated solution.

13.11 A supersaturated solution is a solution that contains more of a solute than is contained in a saturated solution. If you add a crystal to a supersaturated solution, it initiates a crystallization process resulting in a saturated solution, with excess solute settling to the bottom as crystals. In a saturated solution, if you add a crystal, it will simply fall to the bottom, without dissolving.

SOLUTIONS ■ 149

13.12 Since the solubility of potassium dichromate ($K_2Cr_2O_7$) in 100 mL of water is 11.7 g, this is the maximum amount of the material that will dissolve. The excess potassium dichromate that would remain undissolved would be 15.0 g - 11.7 g = 3.3 g.

13.13 In a saturated solution of potassium dichromate, there is an equilibrium that exists where the process of dissolution (in which ions leave the potassium dichromate crystals and enter the solution) and crystallization (in which ions in the solution attach themselves to a crystal). The equilibrium can be represented by

$$K_2Cr_2O_7(s) \rightleftharpoons 2\,K^+\,(aq) + Cr_2O_7^{2-}\,(aq)$$

Even though the amount of potassium dichromate in the solution remains constant, the process of dissolution has not stopped. The rate at which the ions in the solution come back and reattach themselves to crystals becomes equal to the rate at which ions are leaving the crystal. Ions are constantly leaving and returning to the crystals, but at the same rate.

13.14 One reason that some compounds dissolve better than others is due to the attractive forces between the water molecules and the ions. This attractive force is in competition with the force of attraction between ions in the solid. Whichever force is stronger dominates. If the force between ions and water is stronger, the compound dissolves more readily than if the ionic forces are stronger.

13.15

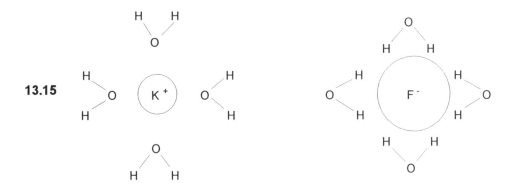

The difference between the two is that for positive ions like K^+, the negative end (the oxygen end) of the polar water molecule is closest to the ion whereas for negative ions like F^-, it is the positive end (the hydrogen end) that is the closest.

13.16 Carbon tetrachloride (CCl_4) is nonpolar. This means that it has no positive or negative end so it will not dissolve in polar liquids. Therefore, it should be more soluble in gasoline, another nonpolar liquid, than in water. The general rule is like dissolves like. Polar substances dissolve in one another and nonpolar substances dissolve in one another, but polar and nonpolar do not mix.

13.17 If you warm up a carbonated beverage, it will go flat. This is because carbon dioxide (carbonation) is more soluble at lower temperatures than at higher temperatures. Thus, as you warm up the beverage, carbon dioxide will bubble out.

13.18 If a saturated solution of sodium nitrate is heated up, it will not remain saturated. Sodium nitrate is more soluble at higher temperatures, so as you heat up the solution, you must dissolve more of the sodium nitrate in order to achieve a saturated solution again. In general, solubility increases with temperature.

13.19 The solubility of a gas, like nitrogen, increases with increasing pressure. Since the pressure exerted on a diver is higher at the deeper depths, more nitrogen dissolves in the blood at the deeper depths. As the diver ascends, the pressure decreases, and the extra nitrogen begins to bubble out. If it occurs too rapidly, it causes the bends.

13.20a. A 4.0% by mass aqueous solution means that for every 100 g of the solution there is 4.0 g of the solute and 96.0 g of water.

    b. A concentrated, 14.5 M ammonia solution has 14.5 moles of ammonia for every liter of solution.

    c. An aqueous sucrose solution that has a concentration of 0.15 m means that there is 0.15 moles of sucrose dissolved in one kilogram of water.

13.21 A conversion factor that converts from liters of solution to moles H_2SO_4 would be

$$\frac{0.18 \text{ mol } H_2SO_4}{1 \text{ liter } H_2SO_4 \text{ solution}}$$

13.22 A conversion factor that converts from kilograms solvent to moles of H_2SO_4 would be

$$\frac{0.18 \text{ mol } H_2SO_4}{1 \text{ kg water}}$$

13.23 Molarity is convenient when you dispense molar amounts of substances by measuring out volumes of solutions, and measuring volumes is simple and very accurate.

13.24 As you increase the temperature of a solution, its volume increases. However, the moles of solute remains constant. This causes a decrease in the molarity (mol/L) of the solution as the temperature increases. On the other hand, the mass of the solvent is not a function of the temperature so the molality (mol/kg) remains constant.

13.25 A titration would be carried out as follows. A measured volume of citric acid solution would be placed in a flask and an indicator is added to indicate the completion of the reaction. A sodium hydroxide (NaOH) solution of known molarity would then be slowly added to the acid solution. When the indicator changes color, the reaction is complete and the volume of the sodium hydroxide solution that was added can now be measured.

The calculation of the mass of citric acid could be determined as follows. First, use the molarity formula to convert liters of NaOH solution to moles of NaOH. Next, use the balanced chemical equation to convert moles of NaOH to moles of citric acid. Finally, multiply the moles of citric acid by the molar mass of citric acid to find the mass of citric acid in the solution.

13.26 To prepare a solution of a specific desired molarity from a more concentrated solution of known molarity, you can use either of the following two methods to determine the volume of the concentrated solution that is needed. The first method is the conversion factor method and the second method is using the dilution equation, $M_iV_i = M_fV_f$. In either case, the volume of concentrated solution is determined. Sufficient water is then added to obtain the final volume of solution.

13.27 A 1.0 L solution of 2.0 M HCl contains 2.0 moles of HCl. Since a 1.0 M solution of HCl can only contain 1.0 mole of HCl per liter, it would take two liters of solution to contain the 2.0 moles of HCl. Therefore, the final volume of the solution would be 2.0 L.

13.28 The freezing point depression, ΔT_f, is the colligative property of a solution equal to the freezing point of the pure solvent minus the freezing point of the solution.

$$\Delta T_f = \text{f.p. pure solvent} - \text{f.p. solution}$$

It is also equal to the freezing point depression constant for the solvent, K_f, multiplied by the molality (for a nonvolatile, nonionic solid). For water, $K_f = 1.86°C/m$

$$\Delta T_f = K_f m$$

Therefore, when m = 1.0, ΔT_f = 1.9°C, and when m = 2.0, ΔT_f = 2.0 x 1.86°C, or 3.7°C.

13.29 For an aqueous solution, $K_f = 1.86°C/m$. For a solution of either glucose or sucrose with concentration of 0.10 m, the freezing point depression is

$$\Delta T_f = K_f m = (1.86°C/m) \times 0.10\ m = 0.19°C$$

Thus, the freezing point of the solution would be –0.19°C (pure water freezes at 0.00°C).

(continued)

K_b for water is 0.52°C/m. The boiling point elevation of a solution of either glucose or sucrose with the same concentration is

$$\Delta T_b = K_b m = (0.52°C/m) \times 0.10\ m = 0.052°C$$

Thus, the boiling point of the solution would be 100.052°C (the boiling point of pure water is 100.000°C).

13.30 Since NaCl forms ions in solution and urea doesn't, there are more solute particles in a NaCl solution than in a solution of urea of the same molality. Since the freezing point depression is a function of the number of solute particles in the solution, the effect is greater for NaCl.

13.31 By the process of osmosis, water would flow out of the greens into the vinegar solution until the concentration of the solute particles is diluted. The effect is that the greens will lose their moisture and wilt.

13.32 A solution that has an osmotic pressure equal to that of blood plasma is isotonic with blood plasma. An example is a 0.85% by mass solution of sodium chloride. This is called a physiological saline solution. If red blood cells are placed in a more concentrated solution, osmosis would occur and water would pass out of the blood cells and they would shrink.

■ Solutions to Practice Problems

<u>Note on significant figures:</u> **When the final answer is written for the first time, it is written with one nonsignificant digit, and the least significant digit is underlined. The final answer is then rounded to the correct number of significant figures.**

13.33a. $\dfrac{15\ g\ Pb(NO_3)_2}{45\ g\ solution} \times 100\% = 3\underline{3}.3 = 33\%$

b. $\dfrac{13.6\ g\ Na_2SO_4}{125\ g\ solution} \times 100\% = 10.\underline{8}8 = 10.9\%$

13.35 First, calculate the mass of the solution:

$$\text{mass solution} = 2.34 \text{ g} + 255 \text{ g} = 257.34 \text{ g}$$

Then, calculate the mass percent:

$$\frac{2.34 \text{ g Pb(NO}_3)_2}{257.34 \text{ g solution}} \times 100\% = 0.90\underline{9}3 = 0.909\%$$

13.37 Use the equation:

$$\text{mass of solute} = \frac{\text{mass percent of solute} \times \text{mass of solution}}{100\%}$$

Substitute in to get the mass of $CuSO_4$:

$$\frac{5.0\% \times 425 \text{ g}}{100\%} = 2\underline{1}.3 = 21 \text{ g}$$

The mass of water is

$$\text{mass water} = 425 \text{ g} - 21.3 \text{ g} = 40\underline{3}.7 = 404 \text{ g}$$

13.39 Rearrange the mass percent equation to get:

$$\text{mass of solution} = \frac{\text{mass of solute} \times 100\%}{\text{mass percent of solute}}$$

Substitute in to get the mass of the solution:

$$\frac{5.0 \text{ g} \times 100\%}{7.35\%} = 6\underline{8}.0 = 68 \text{ g}$$

154 ■ CHAPTER 13

13.41 First, find the mass of NaOH (molecular weight 40.00 amu). So, one mole of NaOH weighs 40.00 g.

$$0.145 \text{ mol NaOH} \times \frac{40.00 \text{ g NaOH}}{1 \text{ mol NaOH}} = 5.800 \text{ g NaOH}$$

Next, find the mass of the solution:

mass solution = 5.800 g + 200. g = 205.800 g

Now, find the mass percent of NaOH:

$$\frac{5.800 \text{ g NaOH}}{205.800 \text{ g solution}} \times 100\% = 2.8\underline{1}8 = 2.82\%$$

13.43a. 250. mL is 0.250 L. So, the molarity is:

$$\frac{0.0450 \text{ mol Cu}(NO_3)_2}{0.250 \text{ L}} = 0.18\underline{0}0 = 0.180 \text{ M}$$

b. 455 mL is 0.455 L. So, the molarity is:

$$\frac{0.250 \text{ mol NH}_4NO_3}{0.455 \text{ L}} = 0.54\underline{9}4 = 0.549 \text{ M}$$

13.45a. First, find the moles of $Ba(NO_3)_2$ (molecular weight 261.32 amu).

$$0.418 \text{ g Ba}(NO_3)_2 \times \frac{1 \text{ mol Ba}(NO_3)_2}{261.32 \text{ g Ba}(NO_3)_2} = 0.001600 \text{ mol Ba}(NO_3)_2$$

Since 25.0 mL is 0.0250 L, the molarity is:

$$\frac{0.001600 \text{ mol Ba}(NO_3)_2}{0.0250 \text{ L}} = 0.063\underline{9}8 = 0.0640 \text{ M}$$

SOLUTIONS ■ 155

b. First, find the moles of NaHSO$_4$ (molecular weight 120.07 amu).

$$23.4 \text{ mg NaHSO}_4 \times \frac{1 \text{ g}}{1000 \text{ mg}} \times \frac{1 \text{ mol NaHSO}_4}{120.07 \text{ g NaHSO}_4} = 1.949 \times 10^{-4} \text{ mol}$$

Since 50.0 mL is 0.0500 L, the molarity is:

$$\frac{1.949 \times 10^{-4} \text{ mol NaHSO}_4}{0.0500 \text{ L}} = 0.003898 = 0.00390 \text{ M}$$

13.47 The molecular weight of K$_2$CrO$_4$ is 194.20 amu. Since 385 mL is 0.385 L, the mass of K$_2$CrO$_4$ needed is

$$0.385 \text{ L solution} \times \frac{0.0750 \text{ mol K}_2\text{CrO}_4}{1 \text{ L solution}} \times \frac{194.20 \text{ g}}{1 \text{ mol K}_2\text{CrO}_4} = 5.607 = 5.61 \text{ g}$$

13.49 $0.150 \text{ mol Na}_2\text{CO}_3 \times \dfrac{1 \text{ L solution}}{0.265 \text{ mol Na}_2\text{CO}_3} \times \dfrac{1000 \text{ mL}}{1 \text{ L}} = 566.0 = 566 \text{ mL}$

13.51 First, find the moles of HCl using the molarity. Since 38.9 mL is 0.0389 L, you get:

$$0.0389 \text{ L solution} \times \frac{0.115 \text{ mol HCl}}{1 \text{ L solution}} = 0.004474 \text{ mol HCl}$$

The molecular weight of CaCO$_3$ is 100.09 amu. Therefore, the mass of CaCO$_3$ in the sample is

$$0.004474 \text{ mol HCl} \times \frac{1 \text{ mol CaCO}_3}{2 \text{ mol HCl}} \times \frac{100.09 \text{ g CaCO}_3}{1 \text{ mol CaCO}_3}$$

$$= 0.2239 = 0.224 \text{ g}$$

13.53 First, find the moles of nitric acid, HNO_3, using the molarity. Since 35.9 mL is 0.0359 L, you get

$$0.0359 \text{ L solution} \times \frac{0.0876 \text{ mol } HNO_3}{1 \text{ L solution}} = 0.003145 \text{ mol } HNO_3$$

The molecular weight of sodium carbonate, Na_2CO_3, is 105.99 amu. Therefore, the mass of Na_2CO_3 required is

$$0.003145 \text{ mol } HNO_3 \times \frac{1 \text{ mol } Na_2CO_3}{2 \text{ mol } HNO_3} \times \frac{105.99 \text{ g } Na_2CO_3}{1 \text{ mol } Na_2CO_3}$$

$$= 0.16\underline{6}7 = 0.167 \text{ g}$$

The molecular weight of $NaNO_3$ is 85.00 amu. Therefore, the mass of $NaNO_3$ produced is

$$0.003145 \text{ mol } HNO_3 \times \frac{2 \text{ mol } NaNO_3}{2 \text{ mol } HNO_3} \times \frac{85.00 \text{ g } NaNO_3}{1 \text{ mol } NaNO_3}$$

$$= 0.26\underline{7}3 = 0.267 \text{ g}$$

13.55 The moles of acetic acid are

$$\frac{0.192 \text{ mol } NaOH}{1 \text{ L solution}} \times 0.0432 \text{ L solution} \times \frac{1 \text{ mol } HC_2H_3O_2}{1 \text{ mol } NaOH}$$

$$= 0.008294 \text{ mol } HC_2H_3O_2$$

The molarity of the acetic acid solution is

$$\text{molarity} = \frac{0.008294 \text{ mol } HC_2H_3O_2}{0.0100 \text{ L solution}} = 0.82\underline{9}4 = 0.829 \text{ M}$$

SOLUTIONS ■ 157

13.57 The equivalents of sulfuric acid that react is

N x V = 0.184 (equiv/L) x 0.0483 L = 0.008887 equiv

The equivalents of Ca(OH)$_2$ initially in the flask is the same. Since 2 equiv Ca(OH)$_2$ equals 1 mol Ca(OH)$_2$, the flask contains 0.008887 ÷ 2 = 0.004444 mol Ca(OH)$_2$. Now, convert this to grams using the molecular weight of Ca(OH)$_2$, 74.10 g/mol. This gives

$$0.004444 \text{ mol} \times \frac{74.10 \text{ g Ca(OH)}_2}{1 \text{ mol Ca(OH)}_2} = 0.3293 = 0.329 \text{ g Ca(OH)}_2$$

13.59 Substitute into the dilution equation. The volume of nitric acid required is

$$\text{volume} = \frac{0.15 \text{ mol/L} \times 0.150 \text{ L}}{15.9 \text{ mol/L}} = 0.00141 = 0.0014 \text{ L} = 1.4 \text{ mL}$$

13.61 Substitute into the dilution equation. The volume of solution needed is

$$\text{volume} = \frac{0.80 \text{ mol/L} \times 0.405 \text{ L}}{2.75 \text{ mol/L}} = 0.118 = 0.12 \text{ L} = 1.2 \times 10^2 \text{ mL}$$

13.63 Substitute into the dilution equation. The molarity of the solution is

$$\text{molarity} = \frac{2.85 \text{ mol/L} \times 0.0500 \text{ L}}{0.250 \text{ L}} = 0.5700 = 0.570 \text{ M}$$

13.65 The molecular weight of C$_2$H$_5$OH is 46.07 amu. The moles of C$_2$H$_5$OH are

$$2.56 \text{ g C}_2\text{H}_5\text{OH} \times \frac{1 \text{ mol C}_2\text{H}_5\text{OH}}{46.07 \text{ g C}_2\text{H}_5\text{OH}} = 0.05557 \text{ mol C}_2\text{H}_5\text{OH}$$

Since 38.6 g is 0.0386 kg, the molality is

$$\frac{0.05557 \text{ mol C}_2\text{H}_5\text{OH}}{0.0386 \text{ kg solvent}} = 1.440 = 1.44 \text{ m}$$

13.67 Solve the molality relationship for kilograms of solvent:

$$\text{mass solvent} = \frac{\text{moles solute}}{\text{molality}} = \frac{0.10 \text{ mol } NH_2CONH_2}{0.34 \text{ mol/kg}}$$

$$= 0.294 \text{ kg} = 294 \text{ g}$$

The mass of urea, NH_2CONH_2, in the solution is (molecular weight 60.06 amu):

$$0.10 \text{ mol } NH_2CONH_2 \times \frac{60.06 \text{ g } NH_2CONH_2}{1 \text{ mol } NH_2CONH_2} = 6.006 \text{ g } NH_2CONH_2$$

The total mass of the solution is 294 g + 6.006 g = 3$\underline{0}$0.006 = 3.0 x 10² g.

13.69 The freezing point depression is

$$\Delta T_f = K_f m = (1.86°C/m) \times (0.25 \text{ m}) = 0.465°C$$

Therefore, the freezing point of the solution is

$$0.00°C - 0.465°C = -0.4\underline{6}5 = -0.47°C$$

13.71 The boiling point elevation of the solution in Problem 13.67 is

$$\Delta T_b = K_b m = (0.52°C/m) \times (0.34 \text{ m}) = 0.177°C$$

Therefore, the boiling point of the solution is

$$100.00°C + 0.177°C = 100.1\underline{7}7 = 100.18°C$$

13.73 Since KBr forms K^+ and Br^- ions in solution, the molality of the solution is doubled, or 2 x 0.015 m = 0.030 m. The freezing point depression is

$$\Delta T_f = K_f m = (1.86°C/m) \times (0.030 \text{ m}) = 0.0558°C$$

The freezing point of the solution is 0.000°C - 0.0558°C = -0.05$\underline{5}$8 = -0.056°C

The boiling point elevation is

$$\Delta T_b = K_b m = (0.52°C/m) \times (0.030 \text{ m}) = 0.0156°C$$

The boiling point of the solution is 100.000°C + 0.0156°C = 100.01$\underline{5}$6 = 100.016°C

SOLUTIONS

13.75 The freezing point depression is

$$\Delta T_f = \text{f. p. pure solvent} - \text{f. p. solution} = 5.455°C - 4.886°C = 0.569°C$$

The molality of the solution is

$$\text{molality} = \frac{\Delta T_f}{K_f} = \frac{0.569\ °C}{5.12\ °C/m} = 0.1111\ m$$

Since 112 g is 0.112 kg, the moles of cholesterol are

$$0.112\ \text{kg solvent} \times \frac{0.1111\ \text{mol cholesterol}}{1\ \text{kg solvent}} = 0.01244\ \text{mol cholesterol}$$

Thus, 4.82 g of cholesterol = 0.01244 mol cholesterol, or

$$\text{molar mass} = \frac{4.82\ g}{0.0124\ \text{mol}} = 38\underline{7}.2 = 387\ g/mol,\ \text{or } 387\ \text{amu}$$

■ Solutions to Additional Problems

13.77 In exactly 100 g of this solution, the mass of solute is

$$\text{mass } NH_4Cl = \frac{9.30\% \times 100\ g}{100\%} = 9.30\ g\ NH_4Cl$$

The molecular weight of NH_4Cl is 53.49 amu. Thus

$$9.30\ g\ NH_4Cl \times \frac{1\ \text{mol } NH_4Cl}{53.49\ g\ NH_4Cl} = 0.1739\ \text{mol } NH_4Cl$$

The mass of water in the solution is

$$\text{mass } H_2O = \text{mass solution} - \text{mass } NH_4Cl = 100\ g - 9.30\ g = 90.70\ g$$

(continued)

This corresponds to 0.09070 kg. The molality of the solution is

$$\text{molality} = \frac{\text{mol NH}_4\text{Cl}}{\text{kg H}_2\text{O}} = \frac{0.1739 \text{ mol}}{0.09070 \text{ kg}} = 1.9\underline{1}7 = 1.92 \text{ m}$$

Since the density of the solution is 1.04 g/mL, the volume of the solution is

$$100 \text{ g} \times \frac{1 \text{ mL}}{1.04 \text{ g}} = 96.15 \text{ mL} = 0.09615 \text{ L}$$

Thus, the molarity of the solution is

$$\text{molarity} = \frac{\text{mol NH}_4\text{Cl}}{\text{L solution}} = \frac{0.1739 \text{ mol}}{0.09615 \text{ L}} = 1.8\underline{0}9 = 1.81 \text{ M}$$

13.79 In two aqueous solutions, the lower freezing point will correspond to the solution with the higher effective molality. Since $BaCl_2$ dissolves in water to form one Ba^{2+} ion and two Cl^- ions, the effective molality is 3 x 0.10 m = 0.30 m. KCl dissolves in water to form one K^+ ion and one Cl^- ion. The effective molality is 2 x 0.10 m = 0.20 m. Since the $BaCl_2$ solution has the bigger effective molality, it will have the lower freezing point.

13.81 Solve the freezing point depression equation for the molality. K_f for water is 1.86 °C/m, and ΔT_f is 2.78°C. Therefore,

$$\text{molality} = \frac{\Delta T_f}{K_f} = \frac{2.78 \, °C}{1.86 \, °C/m} = 1.4946 \text{ m}$$

This is the molality of ions in the solution. Since NaOCl dissociates into two ions, the molality of NaOCl in the solution is one-half of the molality of the ions, or

½(1.4946 m) = 0.74$\underline{7}$3 = 0.747 m.

In one kilogram of solution (1000 g), there are 0.7473 moles of solute. The molecular weight of NaOCl is 74.44 amu. Thus, the mass of NaOCl in the solution is

$$0.7473 \text{ mol} \times \frac{74.44 \text{ g NaOCl}}{1 \text{ mol NaOCl}} = 55.63 \text{ g NaOCl}$$

(continued)

SOLUTIONS ■ 161

The total mass of the solution is

mass solution = mass solute + mass solvent = 55.63g + 1000 g = 1055.63 g

The mass percent of NaOCl is

$$\% \text{ mass} = \frac{55.63 \text{ g NaOCl}}{1055.63 \text{ g solution}} \times 100\% = 5.2\underline{7}0 = 5.27\%$$

13.83 First, use the freezing point depression equation to calculate the molality of the solution. K_f for water is 1.86 °C/m, and ΔT_f is 4.36°C. Therefore,

$$\text{molality} = \frac{\Delta T_f}{K_f} = \frac{4.36 \, °C}{1.86 \, °C/m} = 2.344 \text{ m}$$

Next, use the boiling point elevation equation to calculate ΔT_b. K_b for water is 0.52 °C/m. Thus,

$$\Delta T_b = K_b m = (0.52 \, °C/m) \times (2.344 \text{ m}) = 1.22 \, °C$$

Therefore, the boiling point of the solution is

boiling point = 100.00°C + 1.22°C = 101.$\underline{2}$2 = 101.2°C

13.85 The molarity of the phosphoric acid solution is

$$\text{molarity} = \frac{0.0152 \text{ mol Ba(OH)}_2}{1 \text{ L solution}} \times 0.0347 \text{ L solution} \times \frac{2 \text{ mol H}_3PO_4}{3 \text{ mol Ba(OH)}_2}$$

$$\times \frac{1}{0.00250 \text{ L solution}} = 0.14\underline{0}7 = 0.141 \text{ M}$$

The molecular weight of $Ba_3(PO_4)_2$ is 601.93 amu. The mass of the compound that is formed is

$$\text{mass } Ba_3(PO_4)_2 = \frac{0.0152 \text{ mol Ba(OH)}_2}{1 \text{ L solution}} \times 0.0347 \text{ L} \times \frac{1 \text{ mol Ba}_3(PO_4)_2}{3 \text{ mol Ba(OH)}_2}$$

$$\times \frac{601.93 \text{ g Ba}_3(PO_4)_2}{1 \text{ mol Ba}_3(PO_4)_2} = 0.10\underline{5}8 = 0.106 \text{ g}$$

13.87 The amount of the HCl solution that must be added is

$$\text{volume HCl} = \frac{0.0145 \text{ mol Ca(OH)}_2}{1 \text{ L solution}} \times 1.00 \text{ L solution} \times \frac{2 \text{ mol HCl}}{1 \text{ mol Ca(OH)}_2}$$

$$\times \frac{1 \text{ L solution}}{2.50 \text{ mol HCl}} = 0.01\underline{6}0 = 0.0116 \text{ L}$$

Thus, 0.0116 L, or 11.6 mL of the HCl solution must be added.

13.89 First, determine the total number of moles of HCl that are in the final solution. For the 3.00 M solution, the moles of HCl are

$$\text{mol HCl} = \frac{3.00 \text{ mol HCl}}{1 \text{ L solution}} \times 0.148 \text{ L solution} = 0.4440 \text{ mol}$$

For the 6.25 M solution, the moles of HCl are

$$\text{mol HCl} = \frac{6.25 \text{ mol HCl}}{1 \text{ L solution}} \times 0.252 \text{ L solution} = 1.575 \text{ mol}$$

The total moles of HCl are 0.4440 mol + 1.575 mol = 2.019 mol. The total volume of the final solution is 148 mL + 252 mL = 400. mL, or 0.400 L. The molarity of the mixture is

$$\text{molarity} = \frac{2.019 \text{ mol HCl}}{0.400 \text{ L solution}} = 5.0\underline{4}8 = 5.05 \text{ M}$$

13.91 The molarity of the solution can be determined as follows. In exactly 100 g of the solution, there are 10. g of Cu and 90. g of Ag. Silver is present in the largest amount, so it is the solvent. Thus, there is 0.090 kg of solvent in the solution. Next, find the moles of copper, atomic weight 63.55 amu. Thus,

$$\text{mol Cu} = 10. \text{ g Cu} \times \frac{1 \text{ mol Cu}}{63.55 \text{ g Cu}} = 0.157 \text{ mol}$$

The molality of the solution is

$$\text{molality} = \frac{0.157 \text{ mol}}{0.090 \text{ kg}} = 1.75 \text{ m}$$

(continued)

ΔT_f for this solution is 961°C - 890°C = 71°C. Therefore, the freezing point depression constant for silver is

$$K_f = \frac{\Delta T_f}{m} = \frac{71°C}{1.75\ m} = 40.6 = 41\ °C/m$$

■ Solutions to Practice in Problem Analysis

13.1 To determine the mass percent of solute, you need the mass of the solute, magnesium hydroxide $Mg(OH)_2$, (7.8 g), and the total mass of the solution. The total mass of the solution is calculated by adding the masses of $Mg(OH)_2$ and water (87.6 g). The mass percent is then calculated by dividing the mass of the solute by the mass of the solution, and multiplying by 100 to change into percent.

13.2 To solve this problem, you first would determine the moles of H_2SO_4 in the flask. This is obtained by multiplying the molarity (0.23 M) by the volume of H_2SO_4, in liters (0.035 L). Next, the moles of magnesium hydroxide, $Mg(OH)_2$, are determined by noting that the mole ratio of $Mg(OH)_2$ to H_2SO_4 is 1 to 1. Thus, the moles of $Mg(OH)_2$ and moles of H_2SO_4 are equal. Finally, to determine the mass of $Mg(OH)_2$ in the unknown solution, multiply by the molar mass of $Mg(OH)_2$, 58.33 g/mol.

■ Answers to the Practice Exam

1. d   2. c   3. b   4. d   5. a   6. b   7. a   8. c   9. b   10. e

11. b   12. b   13. a   14. c   15. d

14. REACTION RATES AND CHEMICAL EQUILIBRIUM

■ Solutions to Exercises

Note on significant figures: When the final answer is written for the first time, it is written with one nonsignificant digit, and the least significant digit is underlined. The final answer is then rounded to the correct number of significant figures.

14.1 The balanced equation is

$$4\ NH_3(g) + 3\ O_2(g) \rightleftharpoons 2\ N_2(g) + 6\ H_2O(g)$$

The equilibrium expression is

$$K = \frac{[N_2]^2\ [H_2O]^6}{[NH_3]^4\ [O_2]^3}$$

14.2 The balanced equation is

$$CH_4(g) + 2\ H_2S(g) \rightleftharpoons CS_2(g) + 4\ H_2(g)$$

The equilibrium expression is

$$K = \frac{[CS_2]\ [H_2]^4}{[CH_4]\ [H_2S]^2} = \frac{(1.10)\ (1.68)^4}{(1.10)\ (1.49)^2} = 3.5\underline{5}8 = 3.59$$

14.3 The equilibrium expression is

$$K = \frac{[CH_3OH]}{[CO][H_2]^2} = 2.3$$

Because the equilibrium constant is neither large nor small (around 1), neither the left side nor the right side is favored. Therefore, the reaction is not expected to go to completion at this temperature.

14.4 The equilibrium expression is

$$K = \frac{[CH_4][H_2O]}{[CO][H_2]^3}$$

After solving for [CH$_4$], you obtain

$$[CH_4] = \frac{K \times [CO][H_2]^3}{[H_2O]}$$

Insert the given values to get

$$[CH_4] = \frac{(3.9)(0.30)(0.10)^3}{(0.020)} = 0.5\underline{8}5 = 0.59 \text{ M}$$

14.5 Both Cu(OH)$_2$ and Zn(OH)$_2$ produce the same number of ions when they dissolve. Since K_{sp} for Zn(OH)$_2$ is larger than K_{sp} for Cu(OH)$_2$, Zn(OH)$_2$ is more soluble.

14.6 Since the concentration of H$_2$ is increased, the reaction will shift from left to right (away from H$_2$) in an attempt to remove some of the extra hydrogen.

14.7 Since the volume decreases, the pressure will increase and the equilibrium will shift towards fewer molecules. Thus, the equilibrium will shift from left to right.

14.8 Decreasing the temperature will cause the reaction to shift from right to left, towards the heat. Thus, the concentration of carbon dioxide will increase.

■ Answers to Questions to Test Your Reading

14.1 The factors that influence the rate of a reaction are: (1) the identity of the reactants, (2) the concentration of the reactants, (3) the temperature of the reaction mixture, and (4) the presence of a catalyst.

14.2 In order to react with each other, two molecules must collide. Also, the energy of the collision must be greater than a certain minimum energy, the activation energy, for the molecules to react.

14.3 According to collision theory, a collision between two molecules does not always result in a reaction. There is a minimum energy, the activation energy, that is required before reaction can occur. Not every collision will supply enough energy to break the chemical bonds necessary to cause a reaction. In addition, the molecules must be oriented correctly, or reaction will not occur.

14.4 If the concentration of a reactant is increased, this usually results in an increase in the rate of the reaction.

14.5 Reaction rates increase when the concentrations of the reactants are increased. This is because more molecules will exist in a given volume and thus, more collisions will occur.

14.6 Reaction rates increase with increasing temperature. This is because molecular energies and speeds increase, and thus the average collisional energy will also increase. The likelihood that a given collision will exceed the activation energy will also increase.

14.7 A catalyst is a substance that increases the rate of a reaction without being consumed in the reaction. It provides an alternate but faster pathway for a reaction to occur. The reaction is faster because it has a lower activation energy that the uncatalyzed reaction.

14.8 The activation energy of an uncatalyzed reaction is higher than the activation energy of the same reaction in the presence of a catalyst. The catalyst lowers the activation energy.

14.9 Chemical Equilibrium is a dynamic state in which the rates of the forward and reverse reactions are equal. A chemical system at equilibrium is not a static one because reactants are still being transformed into products and products are still reacting to become the original reactants.

14.10 The law of mass action states that each reaction has an equilibrium constant, K, with its own characteristic value at a given temperature. For the reaction

$$aA + bB \rightleftharpoons cC + dD$$

The equilibrium constant is given by the following ratio of reactants to products.

$$K = \frac{[C]^c [D]^d}{[A]^a [B]^b}$$

where [A], [B], [C], and [D] are the concentrations (in mol/L) of the reactants and products at equilibrium. The equilibrium constant is a constant at a given temperature. If the temperature changes, so does K. Therefore, it is not always constant.

14.11 As long as the temperature is constant, there are an infinite number of equilibrium compositions, each set obtained from different initial concentrations, that correspond to the same equilibrium constant, K.

14.12 The magnitude of the equilibrium constant tells us which side of a chemical equation a particular reaction favors. If the equilibrium constant is large (say, 10^2 or greater), the reaction significantly favors the right side of the chemical equation. Very small values of K (say about 10^{-2} or less), the reaction significantly favors the left side of the chemical equation. When the equilibrium constant is neither large nor small (around 1), neither the left side nor the right side of the chemical equation is favored.

14.13 Homogeneous equilibria are reactions in which all substances are in a single state of matter. Heterogeneous equilibria are reactions in which more than one state of matter is involved.

14.14 a. This reaction has two phases, solid, and aqueous. Thus, it is a heterogeneous equilibrium.

b. This reaction has only one phase, gas. Thus, it is a homogeneous equilibrium.

14.15 Le Chatelier's Principle states that when a reaction at equilibrium is disturbed by a change in volume, temperature, or the concentration of one of the components, the reaction will shift in a way that tends to counteract the change and re-establish the equilibrium.

14.16 You can alter the equilibrium concentrations of a reaction mixture in the following ways:

i. Changing the concentrations by adding or removing substances that appear on either side of the chemical equation (and in the equilibrium expression)

ii. Changing the volume of the system

iii. Changing the temperature

14.17 An exothermic reaction liberates heat. An endothermic reaction absorbs heat from the surroundings.

14.18 For an exothermic reaction at equilibrium, when the temperature is increased, the reaction will shift from right to left in an attempt to absorb the heat and counteract the temperature increase. The equilibrium concentrations of the reactants will increase, and of the products will decrease.

14.19 For an endothermic reaction at equilibrium, when the temperature is increased, the reaction will shift from left to right in an attempt to absorb the heat and counteract the temperature increase. The equilibrium concentrations of the reactants will decrease, and of the products will increase.

14.20 A catalyst has no effect on the equilibrium composition of a reaction; a catalyst merely speeds up the rate at which equilibrium is reached.

14.21 Platinum serves as a catalyst in the reaction. It increases the rate of the reaction. It has no effect on the equilibrium composition of the reaction mixture.

14.22 Four ways to increase the yield of ammonia are

i. Add N_2

ii. Remove NH_3

iii. Decrease the volume

iv. Decrease the temperature

Solutions to Practice Problems

Note on significant figures: When the final answer is written for the first time, it is written with one nonsignificant digit, and the least significant digit is underlined. The final answer is then rounded to the correct number of significant figures.

14.23a. $2 SO_2(g) + O_2(g) \rightleftharpoons 2 SO_3(g)$ $\qquad K = \dfrac{[SO_3]^2}{[SO_2]^2 [O_2]}$

b. $2 NOCl(g) \rightleftharpoons 2 NO(g) + Cl_2(g)$ $\qquad K = \dfrac{[NO]^2 [Cl_2]}{[NOCl]^2}$

c. $POCl_3(g) \rightleftharpoons POCl(g) + Cl_2(g)$ $\qquad K = \dfrac{[POCl][Cl_2]}{[POCl_3]}$

d. $PCl_3(g) + Cl_2(g) \rightleftharpoons PCl_5(g)$ $\qquad K = \dfrac{[PCl_5]}{[PCl_3][Cl_2]}$

14.25a. $CS_2(g) + 4 H_2(g) \rightleftharpoons CH_4(g) + 2 H_2S(g)$ $\qquad K = \dfrac{[CH_4][H_2S]^2}{[CS_2][H_2]^4}$

b. $I_2(g) + Br_2(g) \rightleftharpoons 2 IBr(g)$ $\qquad K = \dfrac{[IBr]^2}{[I_2][Br_2]}$

c. $COCl_2(g) \rightleftharpoons CO(g) + Cl_2(g)$ $\qquad K = \dfrac{[CO][Cl_2]}{[COCl_2]}$

d. $NH_4^+(aq) \rightleftharpoons NH_3(aq) + H^+(aq)$ $\qquad K = \dfrac{[NH_3][H^+]}{[NH_4^+]}$

14.27 $CO(g) + 3 H_2(g) \rightleftharpoons CH_4(g) + H_2O(g)$

$$K = \dfrac{[CH_4][H_2O]}{[CO][H_2]^3} = \dfrac{(0.0387)(0.0387)}{(0.0613)(0.184)^3} = 3.9\underline{2}2 = 3.92$$

14.29 $H_2(g) + I_2(g) \rightleftharpoons 2 HI(g)$

$$K = \dfrac{[HI]^2}{[H_2][I_2]} = \dfrac{(0.0157)^2}{(0.00213)(0.00213)} = 54.\underline{3}3 = 54.3$$

14.31 The equilibrium expression is

$$K = \dfrac{[H_2O]^2}{[H_2]^2[O_2]} = 3 \times 10^{81}$$

The magnitude of the equilibrium constant is very large, so the right side of the equation is favored. Therefore, the reaction will go to completion.

14.33 The equilibrium expression is

$$K = \frac{[CS_2][H_2]^4}{[CH_4][H_2S]^2}$$

After solving for [CS$_2$], you obtain

$$[CS_2] = \frac{K \times [CH_4][H_2S]^2}{[H_2]^4}$$

Insert the given values to get

$$[CS_2] = \frac{(3.59)(1.10)(1.49)^2}{(1.68)^4} = 1.1\underline{0}1 = 1.10 \text{ M}$$

14.35a. $C(s) + CO_2(g) \rightleftharpoons 2\,CO(g)$ $K = \dfrac{[CO]^2}{[CO_2]}$

b. $FeO(s) + CO(g) \rightleftharpoons Fe(s) + CO_2(g)$ $K = \dfrac{[CO_2]}{[CO]}$

c. $2\,Na_2CO_3(s) + 2\,SO_2(g) + O_2(g) \rightleftharpoons 2\,Na_2SO_4(s) + 2\,CO_2(g)$

$$K = \frac{[CO_2]^2}{[SO_2]^2[O_2]}$$

14.37a. $P_4(s) + 5\,O_2(g) \rightleftharpoons P_4O_{10}(s)$ $K = \dfrac{1}{[O_2]^5}$

b. $2\,H_2O_2(l) \rightleftharpoons 2\,H_2O(l) + O_2(g)$ $K = [O_2]$

c. $PbI_2(s) + Cl_2(g) \rightleftharpoons PbCl_2(s) + I_2(g)$ $K = \dfrac{[I_2]}{[Cl_2]}$

REACTION RATES AND CHEMICAL EQUILIBRIUM ■ 171

14.39 Both PbI_2 and Ag_2CrO_4 produce the same number of ions when they dissolve. Since K_{sp} for PbI_2 is larger than K_{sp} for Ag_2CrO_4, PbI_2 is more soluble.

14.41 Since O_2 has been added, the reaction will shift from left to right (away from O_2) to remove some of the extra oxygen.

14.43 Since the volume decreases, the pressure increases and the equilibrium will shift towards fewer molecules. Thus, the equilibrium shifts from left to right.

14.45 For an exothermic reaction, when the temperature is raised, the reaction shifts from right to left, to absorb the heat being added. Thus, CH_3OH would decompose into CO and H_2.

14.47 The equilibrium constant corresponds to the equation

$$CH_4 + 2 H_2S \rightleftharpoons CS_2 + 4 H_2$$

14.49 The equilibrium constant corresponds to the equation

$$2 NO + 2 H_2 \rightleftharpoons N_2 + 2 H_2O$$

14.51 The equilibrium expression is

$$K = [Ag^+][I^-]$$

The solubility product constant is

$$K = (9.1 \times 10^{-9}) \times (9.1 \times 10^{-9}) = 8.28 \times 10^{-17} = 8.3 \times 10^{-17}$$

14.53 The equilibrium expression is

$$K = [Pb^{2+}][Br^-]^2 = (0.010) \times (0.020)^2 = 4.00 \times 10^{-6} = 4.0 \times 10^{-6}$$

■ Solutions to Additional Problems

14.55 a. If H_2 is removed, the reaction will shift from right to left to replace the H_2 removed.

b. If N_2 is added, the reaction will shift from right to left to replace the N_2 added.

c. If the volume is decreased, the pressure is increased and the reaction will shift towards fewer molecules. Thus, the reaction shifts from left to right.

d. If the temperature is decreased, for an exothermic reaction, the reaction shifts from left to right.

e. If a catalyst is added, there will be no change in the equilibrium composition.

14.57 For parts a, b, c, and e, there is no effect on K. For part d, K increases.

14.59a. If Cl⁻ is removed, the reaction will shift from left to right. More $PbCl_2$(s) will dissolve.

b. If $Pb(NO_3)_2$ is added, the reaction will shift from right to left. $PbCl_2$(s) will form.

c. For an endothermic process, when the temperature is increased, the reaction shifts from left to right. More $PbCl_2$ will dissolve.

14.61 For parts a and b, there is no effect on K. For part c, K will increase.

■ Solutions to Practice in Problem Analysis

14.1 To solve this problem, you must determine which direction the reaction will shift in order to establish equilibrium. The value of K (0.500) at this temperature is close to 1, neither the products nor the reactants are favored in the reaction. Some calculations are necessary to determine the initial molarities of the substances. Divide the moles of each substance (1.00 mol N_2, 3.00 mol H_2, and 2.00 mol NH_3) by the volume (50.0 L) to get 0.020 M N_2, 0.060 M H_2, and 0.040 M NH_3. Assume that this mixture represents an equilibrium mixture and calculate the value of K. If the calculated value is larger than the actual value of K, then there is too much of the substances on the right hand side of the equation. The reaction will shift from right to left, reducing the amount of NH_3 present at equilibrium. If it is too small, then there is not enough of the substances on the right hand side. The reaction will shift from left to right, and the amount of NH_3 present at equilibrium will increase.

14.2 To solve this problem, use Le Chatelier's Principle to determine the direction that the reaction shifts to reestablish equilibrium after the increase in temperature. Since the amount of CH_3OH decreases upon heating, the reaction shifts from right to left. An increase in temperature also causes the reaction to shift away from the heat. This means that the heat is on the same side as the CH_3OH, so the reaction liberates heat making it exothermic.

Answers to the Practice Exam

1. a   2. e   3. a   4. c   5. d   6. a   7. c   8. c   9. b   10. b
11. c   12. c   13. b   14. a   15. a

15. ACIDS AND BASES

■ Solutions to Exercises

Note on significant figures: When the final answer is written for the first time, it is written with one nonsignificant digit, and the least significant digit is underlined. The final answer is then rounded to the correct number of significant figures.

15.1 Write the formulas for the acid, the base, and the salt as separate ions to get the total ionic equation.

$$Ba^{2+}(aq) + 2\,OH^-(aq) + 2\,H^+(aq) + 2\,Cl^-(aq) \longrightarrow Ba^{2+}(aq) + 2\,Cl^-(aq) + 2\,H_2O(l)$$

Barium ions and chloride ions are spectator ions because they have not changed. They can be eliminated. Also, you can cancel the 2 in front of the hydrogen and hydroxide ions, and in front of water. The result is the net ionic equation.

$$H^+(aq) + OH^-(aq) \longrightarrow H_2O(l)$$

15.2 The conjugate acid of OH^- can be obtained by adding an H^+ ion to it. Therefore, H_2O is the conjugate acid of OH^-.

15.3 Examine the equation to determine the H^+ donor on each side. On the left, H_2CO_3 is the donor, and on the right, HCN is the donor. These substances are the acids. The acceptors are CN^- and HCO_3^-, and these are the bases. This gives

$$H_2CO_3(aq) + CN^-(aq) \rightleftharpoons HCN(aq) + HCO_3^-(aq)$$
$$\text{Acid} \qquad\qquad \text{Base} \qquad\qquad \text{Acid} \qquad\qquad \text{Base}$$

(continued)

ACIDS AND BASES ■ 175

Next, identify the pairs of substances that differ by the gain or loss of a proton.

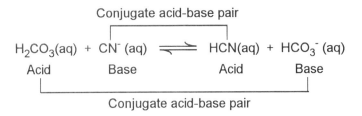

Thus, H_2CO_3 and HCO_3^- are a conjugate acid-base pair, as are HCN and CN^-.

15.4 With $[H_3O^+] = 8.4 \times 10^{-5}$ M, obtain the concentration of hydroxide ions from the expression for K_W.

$$K_W = [H_3O^+][OH^-]$$

$$1.0 \times 10^{-14} = (8.4 \times 10^{-5}) \times [OH^-]$$

$$[OH^-] = \frac{1.0 \times 10^{-14}}{8.4 \times 10^{-5}} = 1.\underline{1}9 \times 10^{-10} = 1.2 \times 10^{-10} \text{ M}$$

15.5 Since 4.7×10^{-5} M is greater than 1.0×10^{-7} M, the solution is acidic.

15.6 On a calculator, do the following operations:

1. Press 2.2
2. Press EXP (or EE)
3. Press 11
4. Press +/-
5. Press LOG
6. Press +/-

The result is 10.6$\underline{5}$8. Thus, pH = 10.66.

15.7 Do the following operations on your calculator.

1. Press 5
2. Press EXP (or EE)
3. Press 8
4. Press +/-
5. Press LOG
6. Press +/-

The result is 7.30. Thus, the pH = 7.3, and the solution is basic.

15.8 Using a calculator, enter the following operations

1. Press 3.16
2. Press +/-
3. Press INV
4. Press LOG

The result is 6.92 x 10^{-4}. Thus,

$[H_3O^+]$ = 6.9 x 10^{-4} M

The solution is acidic.

■ Answers to Questions to Test Your Reading

15.1 According to Arrhenius, an acid is a substance that produces H$^+$ ions (protons) when it is dissolved in water. An example is nitric acid, HNO$_3$. A base is a substance that produce OH$^-$ ions when it is dissolved in water. An example is sodium hydroxide, NaOH.

15.2 The net ionic equation that describes the reaction between HBr and KOH is

H$^+$(aq) + OH$^-$(aq) $\longrightarrow$ H$_2$O(l)

15.3 A salt is an ionic compound containing the cation from a base and an anion from an acid. It is also one of the products of an acid-base neutralization reaction. Ordinary table salt, which is sodium chloride (NaCl), fits this description. The cation (Na$^+$) comes from the base NaOH, and the anion (Cl$^-$) comes from the acid HCl.

15.4 According to the Bronsted-Lowry concept, an acid is a proton (H^+) donor and a base is a proton acceptor. Examples of an acid and a base are hydrofluoric acid (HF) and ammonia (NH_3).

15.5 Hydronium ion is H_3O^+ and has the structure

$$H : \overset{..}{\underset{..}{\underset{|}{O}}} : H \quad +$$
$$\phantom{H : \overset{..}{\underset{..}{O}}}H$$

15.6 A conjugate acid-base pair consists of two substances (one an acid and the other a base) in an acid-base reaction that differ by the gain or loss of a proton. For example

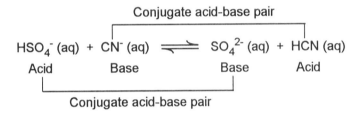

The HSO_4^- and the SO_4^{2-} ions are a conjugate acid-base pair, as are HCN and CN^-.

15.7 The reaction of an acid with water is represented by

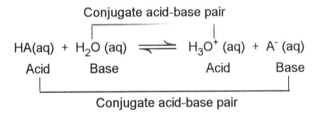

15.8 Water is amphoteric because it can behave as both an acid and a base. This is best seen in the self-ionization of water in which a proton is transferred from one water molecule to another.

$$\underset{\text{Acid}}{H_2O(l)} + \underset{\text{Base}}{H_2O(l)} \rightleftharpoons \underset{\text{Acid}}{H_3O^+(aq)} + \underset{\text{Base}}{OH^-(aq)}$$

Conjugate acid-base pair (H$_2$O / H$_3$O$^+$ and H$_2$O / OH$^-$)

Thus, H$_2$O is the conjugate acid of OH$^-$ and the conjugate base of H$_3$O$^+$.

15.9 A monoprotic acid can only transfer one proton to a base. An example is HCl, hydrochloric acid. A diprotic acid can transfer two protons. An example is H$_2$SO$_4$, sulfuric acid.

15.10 Acids that have the ability to transfer a proton to water that is very great are strong acids. They are completely dissociated in water. They are: hydrochloric acid, HCl; hydrobromic acid, HBr; hydroiodic acid, HI; nitric acid, HNO$_3$; perchloric acid, HClO$_4$; and sulfuric acid, H$_2$SO$_4$. (See Table 15.1)

15.11 Acids that have a poor ability to transfer a proton to water are weak acids. They are not completely dissociated in water. An example is H$_2$SO$_3$, sulfurous acid. (See Table 15.2)

15.12 Since the strength of a conjugate base increases as the strength of an acid decreases, the strongest bases have the weakest conjugate acids. They are totally ionized in water. Four examples are: lithium hydroxide, LiOH; sodium hydroxide, NaOH; potassium hydroxide, KOH; and calcium hydroxide, Ca(OH)$_2$. (See Table 15.1)

15.13 A weak base is a base that is not totally ionized in water. An example is ammonia, NH$_3$.

15.14 The equilibrium expression for the reaction of HNO$_2$ with water is

$$K_a = \frac{[H_3O^+][NO_2^-]}{[HNO_2]}$$

15.15 The stronger the acid, the weaker the conjugate base. Since formic acid, HCHO$_2$, is a stronger acid than acetic acid, HC$_2$H$_3$O$_2$, therefore formate ion, CHO$_2^-$, is a weaker base than acetate ion, C$_2$H$_3$O$_2^-$.

ACIDS AND BASES ■ 179

15.16 Since formic acid is a stronger acid than acetic acid, it has a greater tendency to transfer protons to water. Thus, a 0.1 M solution of formic acid will transfer more protons to water than a 0.1 M solution of acetic acid, and thus have a higher concentration of hydronium ions.

15.17 H_2SO_3 is a stronger acid than HSO_3^- because HSO_3^- is the conjugate base of H_2SO_3.

15.18 Self-ionization means that two identical molecules react to give ions. The self-ionization of water to give H_3O^+ and OH^- ions is an example.

15.19 The ion-product constant for water (K_W) is the equilibrium value of the ion product $[H_3O^+][OH^-]$. At 25 °C, the value is

$$K_W = [H_3O^+][OH^-] = 1.0 \times 10^{-14}$$

15.20 In terms of hydrogen ion concentrations, in an acidic solution the concentration of H^+ ions is greater than that of OH^- ions. In a basic solution, the concentration of OH^- ions is greater than that of H^+ ions. In a neutral solution, the concentrations of H^+ and OH^- ions are equal. At 25°C,

In an acidic solution, $[H_3O^+] > 1.0 \times 10^{-7}$ M

In a neutral solution, $[H_3O^+] = 1.0 \times 10^{-7}$ M

In a basic solution, $[H_3O^+] < 1.0 \times 10^{-7}$ M

15.21 pH is defined as a unitless number obtained from the negative of the logarithm of the hydrogen ion concentrations.

$$pH = -\log[H_3O^+]$$

The pH scale normally runs from 0 to 14. However, for strong acids or bases, pH values below -1 and above 15, respectively, are possible.

In an acidic solution, pH < 7.00

In a neutral solution, pH = 7.00

In a basic solution, pH > 7.00

15.22 Rainwater with a pH of 4.3 would be acidic.

15.23 pH can be measured in the following two ways. First, using a pH meter. Second, using acid-base indicators. pH paper can also be used to measure pH.

180 ■ CHAPTER 15

15.24 An indicator is a weak acid whose solution will change color within a small pH range and indicate the pH of the solution by the color. If HIn represents the acid form of the indicator, and In⁻ its conjugate base, then the equilibrium that exists is given by

$$HIn(aq) + H_2O(l) \rightleftharpoons H_3O^+(aq) + In^-(aq)$$

The acid form HIn is a different color than its conjugate base In⁻. The colors are so intense that only a small quantity is needed, and it does not affect the pH of the test solution. The test solution's hydronium ion concentration will determine the concentrations of acid and base forms of the indicator by causing the indicator reaction to shift until equilibrium is reestablished. Thus, the pH of the test solution will dictate the color of the solution.

15.25 Bromocresol green is yellow below pH 3.8 and blue above pH 5.4. Thus, when a solution that contains bromocresol green is green, the pH is approximately 4-5.

15.26 When bromocresol green is yellow, the pH is less than 3.8. When methyl violet is violet, the pH is above 1.6. Thus the solution has a pH of approximately 2-4.

15.27 A buffer is a solution that can resist changes in pH when limited amounts of acid or base are added to it. They contain a weak acid and its conjugate base. An example is NH_4^+ and NH_3.

15.28 When a strong acid is added to an ammonia-ammonium chloride buffer, H_3O^+ ions will react with the base, NH_3, to form NH_4^+, the conjugate acid. When a strong base is added to the buffer, it reacts with the acid, NH_4^+, to form the conjugate base, NH_3.

■ Solutions to Practice Problems

Note on significant figures: When the final answer is written for the first time, it is written with one nonsignificant digit, and the least significant digit is underlined. The final answer is then rounded to the correct number of significant figures.

15.29 The total ionic equation is

$$Ra^{2+}(aq) + 2\,OH^-(aq) + 2\,H^+(aq) + 2\,Cl^-(aq) \longrightarrow Ra^{2+}(aq) + 2\,Cl^-(aq) + 2\,H_2O(l)$$

Ra^{2+} and Cl^- are spectator ions. After cancelling the two's, the net ionic equation is

$$H^+(aq) + OH^-(aq) \longrightarrow H_2O(l)$$

15.31 $HClO_4(aq) + H_2O(l) \longrightarrow H_3O^+(aq) + ClO_4^-(aq)$

15.33 a. F^- b. HS^- c. S^{2-} d. OH^-

15.35 a. NH_4^+ b. H_2S c. HS^- d. HCN

15.37 a.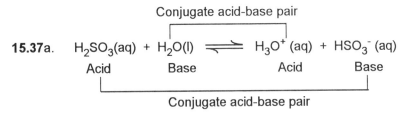

Thus, H_2SO_3 and HSO_3^- are a conjugate acid-base pair, as are H_3O^+ and H_2O.

b.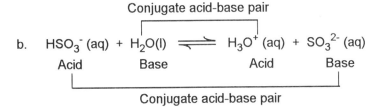

Thus, HSO_3^- and SO_3^{2-} are a conjugate acid-base pair, as are H_3O^+ and H_2O.

c.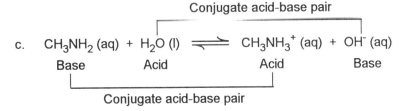

Thus, $CH_3NH_3^+$ and CH_3NH_2 are a conjugate acid-base pair, as are H_2O and OH^-.

d.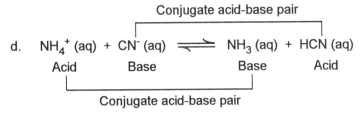

Thus, NH_4^+ and NH_3 are a conjugate acid-base pair, as are HCN and CN^-.

15.39a.

Conjugate acid–base pair

$$CN^- (aq) + H_2O (l) \rightleftharpoons HCN (aq) + OH^- (aq)$$

Base — Acid — Acid — Base

Conjugate acid–base pair

Thus, HCN and CN^- are a conjugate acid-base pair, as are H_2O and OH^-.

b.

$$HC_2H_3O_2 (aq) + OH^- (aq) \rightleftharpoons C_2H_3O_2^- (aq) + H_2O (l)$$

Acid — Base — Base — Acid

Thus, $HC_2H_3O_2$ and $C_2H_3O_2^-$ are a conjugate acid-base pair, as are H_2O and OH^-.

c.

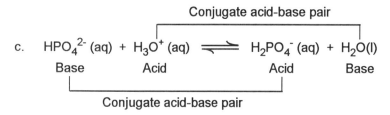

$$HPO_4^{2-} (aq) + H_3O^+ (aq) \rightleftharpoons H_2PO_4^- (aq) + H_2O(l)$$

Base — Acid — Acid — Base

Thus, $H_2PO_4^-$ and HPO_4^{2-} are a conjugate acid-base pair, as are H_3O^+ and H_2O.

d.

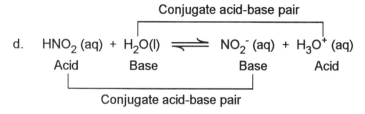

$$HNO_2 (aq) + H_2O(l) \rightleftharpoons NO_2^- (aq) + H_3O^+ (aq)$$

Acid — Base — Base — Acid

Thus, HNO_2 and NO_2^- are a conjugate acid-base pair, as are H_3O^+ and H_2O.

ACIDS AND BASES ■ 183

15.41a. $[OH^-] = \dfrac{K_W}{[H_3O^+]} = \dfrac{1.0 \times 10^{-14}}{0.25} = 4.0\underline{0} \times 10^{-14} = 4.0 \times 10^{-14}$ M

b. $[OH^-] = \dfrac{K_W}{[H_3O^+]} = \dfrac{1.0 \times 10^{-14}}{0.0035} = 2.8\underline{6} \times 10^{-12} = 2.9 \times 10^{-12}$ M

c. $[OH^-] = \dfrac{K_W}{[H_3O^+]} = \dfrac{1.0 \times 10^{-14}}{1.2 \times 10^{-2}} = 8.3\underline{3} \times 10^{-13} = 8.3 \times 10^{-13}$ M

d. $[OH^-] = \dfrac{K_W}{[H_3O^+]} = \dfrac{1.0 \times 10^{-14}}{4.2 \times 10^{-11}} = 2.3\underline{8} \times 10^{-4} = 2.4 \times 10^{-4}$ M

15.43 $[OH^-] = \dfrac{K_W}{[H_3O^+]} = \dfrac{1.0 \times 10^{-14}}{3.7 \times 10^{-4}} = 2.\underline{70} \times 10^{-11} = 2.7 \times 10^{-11}$ M

15.45a. $[H_3O^+] = 5 \times 10^{-9}$ M, and is less than 1.0×10^{-7} M. Therefore, the solution is basic.

b. First, obtain the concentration of H_3O^+ ions using the expression for K_W.

$[H_3O^+] = \dfrac{K_W}{[OH^-]} = \dfrac{1.0 \times 10^{-14}}{1.0 \times 10^{-7}} = 1.0\underline{0} \times 10^{-7} = 1.0 \times 10^{-7}$ M

Thus, the solution is neutral.

c. First, obtain the concentration of H_3O^+ ions using the expression for K_W.

$[H_3O^+] = \dfrac{K_W}{[OH^-]} = \dfrac{1.0 \times 10^{-14}}{5 \times 10^{-9}} = 2.\underline{0} \times 10^{-6} = 2 \times 10^{-6}$ M

Since 2×10^{-6} M is greater than 1.0×10^{-7} M, the solution is acidic.

d. $[H_3O^+] = 2 \times 10^{-7}$ M, and is greater than 1.0×10^{-7} M. Therefore, the solution is acidic.

15.47 First, obtain the concentration of H_3O^+ ions using the expression for K_W.

$$[H_3O^+] = \frac{K_W}{[OH^-]} = \frac{1.0 \times 10^{-14}}{1.5 \times 10^{-7}} = 6.\underline{6}7 \times 10^{-8} = 6.7 \times 10^{-8} \text{ M}$$

Since 6.7×10^{-8} M is less than 1.0×10^{-7} M, the solution is basic.

15.49 a. 8.00 b. 2.12 c. 11.30 d. 8.20

15.51 The pH of the vinegar, $[H_3O^+] = 7.5 \times 10^{-3}$ M, is 2.12. Since 2.12 is less than 7.00, the solution is acidic.

15.53 a. First, calculate $[H_3O^+]$ using K_W and $[OH^-]$.

$$[H_3O^+] = \frac{K_W}{[OH^-]} = \frac{1.0 \times 10^{-14}}{1.0 \times 10^{-8}} = 1.00 \times 10^{-6} \text{ M}$$

Next, calculate the pH.

$$pH = -\log(1.00 \times 10^{-6}) = 6.0\underline{0}0 = 6.00$$

b. First, calculate $[H_3O^+]$ using K_W and $[OH^-]$.

$$[H_3O^+] = \frac{K_W}{[OH^-]} = \frac{1.0 \times 10^{-14}}{7.5 \times 10^{-3}} = 1.33 \times 10^{-12} \text{ M}$$

Next, calculate the pH.

$$pH = -\log(1.33 \times 10^{-12}) = 11.8\underline{7}6 = 11.88$$

c. First, calculate $[H_3O^+]$ using K_W and $[OH^-]$.

$$[H_3O^+] = \frac{K_W}{[OH^-]} = \frac{1.0 \times 10^{-14}}{5.0 \times 10^{-12}} = 2.00 \times 10^{-3} \text{ M}$$

Next, calculate the pH.

$$pH = -\log(2.00 \times 10^{-3}) = 2.6\underline{9}9 = 2.70$$

ACIDS AND BASES

d. First, calculate $[H_3O^+]$ using K_W and $[OH^-]$.

$$[H_3O^+] = \frac{K_W}{[OH^-]} = \frac{1.0 \times 10^{-14}}{6.3 \times 10^{-9}} = 1.59 \times 10^{-6} \text{ M}$$

Next, calculate the pH.

$$pH = -\log(1.59 \times 10^{-6}) = 5.7\underline{9}9 = 5.80$$

15.55 First, find the concentration of H_3O^+ ions using the expression for K_W.

$$[H_3O^+] = \frac{K_W}{[OH^-]} = \frac{1.0 \times 10^{-14}}{0.0040} = 2.50 \times 10^{-12} \text{ M}$$

Next, calculate the pH.

$$pH = -\log(2.50 \times 10^{-12}) = 11.6\underline{0}2 = 11.60$$

Since 11.60 is more than 7.00, the solution is basic.

15.57 a. $[H_3O^+]$ = inverse log (-7.28) = 5.2×10^{-8} M.

b. $[H_3O^+]$ = inverse log (-2.32) = 4.8×10^{-3} M.

c. $[H_3O^+]$ = inverse log (-12.10) = 7.9×10^{-13} M.

d. $[H_3O^+]$ = inverse log (-6.73) = 1.9×10^{-7} M.

15.59 $[H_3O^+]$ = inverse log (-8.74) = $1.\underline{8}2 \times 10^{-9}$ = 1.8×10^{-9} M.

The OH⁻ concentration is obtained from the expression for K_W.

$$[OH^-] = \frac{K_W}{[H_3O^+]} = \frac{1.0 \times 10^{-14}}{1.82 \times 10^{-9}} = 5.\underline{4}9 \times 10^{-6} = 5.5 \times 10^{-6} \text{ M}$$

Solutions to Additional Problems

15.61 Since pH is a number represented by a logarithm, a difference of 1 pH unit is the same as 10 raised to the 1 power, or a factor of 10. Thus, the relative difference in the concentration of hydronium ion in the solutions whose pH differs by 1 pH unit is that the solution with the lower pH has a $[H_3O^+]$ 10 times greater than that of the solution with the higher pH. As an example, a solution with pH 2.0 has a $[H_3O^+]$ of 1×10^{-2} M, and one with pH = 3.0 has a $[H_3O^+]$ of 1×10^{-3} M, a difference of a factor of 10.

15.63 First, calculate the moles of NaOH present (molar mass, 40.00 g/mol).

$$40.\text{ g} \times \frac{1 \text{ mol}}{40.00 \text{ g}} = 1.00 \text{ mol}$$

Next, calculate the concentration of NaOH.

$$[NaOH] = \frac{\text{mol NaOH}}{\text{L solution}} = \frac{1.00 \text{ mol}}{10 \text{ L}} = 0.100 \text{ M}$$

Therefore, since NaOH is a strong base, the hydroxide concentration is 0.100 M. Using K_W gives the hydronium ion concentration.

$$[H_3O^+] = \frac{K_W}{[OH^-]} = \frac{1.0 \times 10^{-14}}{0.100} = 1.00 \times 10^{-13} \text{ M}$$

The pH of the solution is

$$\text{pH} = -\log(1.00 \times 10^{-13}) = 13.0\underline{0}0 = 13.00$$

Solutions to Practice in Problem Analysis

15.1 The first step in solving this problem is to determine the hydronium ion concentration using the pH. Next, use the hydronium ion concentration and the relationship for K_W ($K_W = [H_3O^+][OH^-]$) to determine the hydroxide ion concentration. Then, write a balanced equation to determine the mole ratio of NH_4^+ to OH^-. Finally, multiply the concentration of hydronium ion by the mole ratio (1 to 1) to determine the NH_4^+ concentration.

15.2 To solve this problem, first determine the moles of HCl by multiplying the concentration (0.633 M) by the volume used, in liters (0.0456 L). Then, write a balanced equation for the reaction to determine the mole ratio of NaOH to HCl (1 to 1). Multiply the moles of HCl by this ratio. Finally, multiply the moles of NaOH by its molar mass (40.00 g/mol) to get the grams of NaOH.

Answers to the Practice Exam

1. b   2. c   3. c   4. a   5. c   6. a   7. a   8. b   9. c   10. b

11. a   12. c   13. d   14. b   15. d

16. OXIDATION-REDUCTION REACTIONS

■ Solutions to Exercises

16.1 The charge on the nickel ion increases ($Ni^{2+} \rightarrow Ni^{3+}$) so Ni^{2+} is oxidized (loses electrons), and it is the reducing agent. The charge on the cadmium ion decreases ($Cd^{2+} \rightarrow Cd$) so Cd^{2+} is reduced (gains electrons), and it is the oxidizing agent.

16.2 Magnesium is above silver in the activity series. Therefore, a reaction between Mg and Ag^+ will occur.

$$Mg(s) + 2\,Ag^+(aq) \longrightarrow Mg^{2+}(aq) + 2\,Ag(s)$$

16.3 Na_2CrO_4: First, we can get the oxidation number for sodium (Rule 7) and oxygen (Rule 6). This gives

$$\underset{\substack{+1 \\ \text{from Rule 7}}}{Na_2}\,Cr\,\underset{\substack{-2 \\ \text{from Rule 6}}}{O_4}$$

Next, the oxidation number of chromium can be obtained from Rule 8 (the sum of the oxidation numbers for the atoms in a compound is zero). In this case, we have

$$2 \times (+1) + \text{oxidation number for chromium} + 4 \times (-2) = 0$$

$$2 + \text{oxidation number for chromium} - 8 = 0$$

$$\text{oxidation number for chromium} = +6$$

Thus, the oxidation number of chromium in Na_2CrO_4 is +6.

(continued)

$Na_2Cr_2O_7$: First, we can get the oxidation number for sodium (Rule 7) and oxygen (Rule 6). This gives

$$Na_2Cr_2O_7$$
$$\quad +1 \qquad\qquad -2$$
$$\text{from Rule 7} \quad \text{from Rule 6}$$

Next, the oxidation number of chromium can be obtained from Rule 8 (the sum of the oxidation numbers for the atoms in a compound is zero). In this case, we have

$$2 \times (+1) + 2 \times \text{oxidation number of chromium} + 7 \times (-2) = 0$$
$$2 + 2 \times \text{oxidation number of chromium} - 14 = 0$$
$$2 \times \text{oxidation number of chromium} = 12$$
$$\text{oxidation number of chromium} = +6$$

Thus, the oxidation number of chromium in Na_2CrO_4 is +6.

16.4 $S_2O_3^{2-}$: The oxidation number of oxygen is -2 (Rule 6). We can obtain the oxidation number of sulfur from Rule 8 (the sum of the oxidation numbers for the atoms in a polyatomic ion will equal the charge on the ion). Thus,

$$2 \times \text{oxidation number for S} + 3 \times (-2) = -2$$
$$2 \times \text{oxidation number for S} - 6 = -2$$
$$2 \times \text{oxidation number for S} = +4$$
$$\text{oxidation number for S} = +2$$

16.5 First, you need to assign oxidation numbers to each substance.

$$Na_2O(s) + H_2O(l) \longrightarrow 2\ NaOH(aq)$$
$$+1 \quad -2 \quad +1 \quad -2 \qquad\qquad +1 \quad -2 \quad +1$$

Since there is no change in the oxidation number of any atom (Na is +1, H is +1, and O is -2), this is not an oxidation-reduction reaction.

16.6 Use the nine steps as outlined in the text.

1. Assign oxidation numbers to each atom in the equation and decide which atoms are oxidized and which atoms are reduced.

$$\overset{+1\quad -2}{H_2S} + \overset{+5\quad -2}{NO_3^-} \longrightarrow \overset{0}{S_8} + \overset{+2\;-2}{NO}$$

The anion NO_3^- is reduced because the oxidation number of nitrogen decreases; H_2S is oxidized because the oxidation number of sulfur increases. The NO_3^- ion is the oxidizing agent and H_2S is the reducing agent.

2. Split the skeletal equation into a half-reaction for oxidation and a half-reaction for reduction.

$H_2S \longrightarrow S_8$ (Oxidation)

$NO_3^- \longrightarrow NO$ (Reduction)

3. Balance all atoms in each half-reaction except hydrogen and oxygen.

$8\,H_2S \longrightarrow S_8$

$NO_3^- \longrightarrow NO$

4. Balance oxygen by adding H_2O where needed.

$8\,H_2S \longrightarrow S_8$

$NO_3^- \longrightarrow NO + 2\,H_2O$

5. Balance hydrogen by adding H^+ where needed.

$8\,H_2S \longrightarrow S_8 + 16\,H^+$

$NO_3^- + 4\,H^+ \longrightarrow NO + 2\,H_2O$

Since the reaction occurs in acidic solution, you do not need to introduce OH^- ions.

6. Balance charge by adding electrons where needed.

$8\,H_2S \longrightarrow S_8 + 16\,H^+ + 16\,e^-$

$NO_3^- + 4\,H^+ + 3\,e^- \longrightarrow NO + 2\,H_2O$

(continued)

7. Multiply the half-reactions by factors that will lead to the same number of electrons in each half-reaction.

$$3 \times [8\,H_2S \longrightarrow S_8 + 16\,H^+ + 16\,e^-] =$$
$$24\,H_2S \longrightarrow 3\,S_8 + 48\,H^+ + 48\,e^-$$

$$16 \times [NO_3^- + 4\,H^+ + 3\,e^- \longrightarrow NO + 2\,H_2O] =$$
$$16\,NO_3^- + 64\,H^+ + 48\,e^- \longrightarrow 16\,NO + 32\,H_2O$$

8. Add the two half-reactions.

$$24\,H_2S \longrightarrow 3\,S_8 + 48\,H^+ + 48\,e^-$$
$$16\,NO_3^- + 64\,H^+ + 48\,e^- \longrightarrow 16\,NO + 32\,H_2O$$

$$24\,H_2S + 16\,NO_3^- + \underset{16}{\cancel{64}}\,H^+ + \cancel{48\,e^-} \longrightarrow 3\,S_8 + 16\,NO + 32\,H_2O + \cancel{48\,H^+} + \cancel{48\,e^-}$$

Notice that 48 electrons on each side have been canceled. In addition, 48 H^+ ions on the right cancel the same number on the left, causing a decrease from 64 H^+ to 16 H^+. The result is

$$24\,H_2S + 16\,NO_3^- + 16\,H^+ \longrightarrow 3\,S_8 + 16\,NO + 32\,H_2O$$

9. Check the equation to make sure it is balanced.

	Left	Right
H	64	64
S	24	24
N	16	16
O	48	48
Charge	0	0

The balanced equation, including phase labels, is

$$24\,H_2S(aq) + 16\,NO_3^-(aq) + 16\,H^+(aq) \longrightarrow$$
$$3\,S_8(s) + 16\,NO(g) + 32\,H_2O(l)$$

16.7 Follow the nine steps outlined in the text.

1. Assign oxidation numbers to each atom in the equation and decide which atoms are oxidized and which atoms are reduced.

$$\overset{+7\ -2}{MnO_4^-} + \overset{-1}{I^-} \longrightarrow \overset{+4\ -2}{MnO_2} + \overset{0}{I_2}$$

The anion MnO_4^- is reduced because the oxidation number of manganese decreases, and it is the oxidizing agent; I^- is oxidized because the oxidation number of iodine increases, and it is the reducing agent.

2. Split the skeletal equation into a half-reaction for oxidation and a half-reaction for reduction.

$$I^- \longrightarrow I_2 \quad \text{(Oxidation)}$$
$$MnO_4^- \longrightarrow MnO_2 \quad \text{(Reduction)}$$

3. Balance all atoms in each half-reaction except hydrogen and oxygen.

$$2\,I^- \longrightarrow I_2$$
$$MnO_4^- \longrightarrow MnO_2$$

4. Balance oxygen by adding H_2O where needed.

$$2\,I^- \longrightarrow I_2$$
$$MnO_4^- \longrightarrow MnO_2 + 2\,H_2O$$

5. Balance hydrogen by adding H^+ where needed.

$$2\,I^- \longrightarrow I_2$$
$$MnO_4^- + 4\,H^+ \longrightarrow MnO_2 + 2\,H_2O$$

Notice that 4 H^+ occurs in the reduction half-reaction. Since the reaction occurs in basic solution, change the half-reactions from acidic to basic conditions by adding 4 OH^- to each side of the reduction half-reaction.

$$MnO_4^- + 4\,H^+ + 4\,OH^- \longrightarrow MnO_2 + 2\,H_2O + 4\,OH^-$$

(continued)

Combine the 4 H⁺ and 4 OH⁻ on the left side to give 4 H₂O molecules. Cancel two water molecules from each side of the half-reaction.

$$MnO_4^- + 4\,H_2O \longrightarrow MnO_2 + \cancel{2}\,H_2O + 4\,OH^-$$
$$\ \ 2$$

The result is

$$MnO_4^- + 2\,H_2O \longrightarrow MnO_2 + 4\,OH^-$$

6. Balance the charge by adding electrons where needed.

$$2\,I^- \longrightarrow I_2 + 2\,e^-$$
$$MnO_4^- + 2\,H_2O + 3\,e^- \longrightarrow MnO_2 + 4\,OH^-$$

7. Multiply each half-reaction by factors that will lead to the same number of electrons in each half-reaction.

$$3 \times [\,2\,I^- \longrightarrow I_2 + 2\,e^-\,] = 6\,I^- \longrightarrow 3\,I_2 + 6\,e^-$$
$$2 \times [\,MnO_4^- + 2\,H_2O + 3\,e^- \longrightarrow MnO_2 + 4\,OH^-\,] =$$
$$2\,MnO_4^- + 4\,H_2O + 6\,e^- \longrightarrow 2\,MnO_2 + 8\,OH^-$$

8. Add the half-reactions. Cancel six electrons from both sides of the overall equation.

$$6\,I^- \longrightarrow 3\,I_2 + 6\,e^-$$
$$2\,MnO_4^- + 4\,H_2O + 6\,e^- \longrightarrow 2\,MnO_2 + 8\,OH^-$$
$$\overline{}$$
$$2\,MnO_4^- + 6\,I^- + 4\,H_2O + \cancel{6\,e^-} \longrightarrow 2\,MnO_2 + 3\,I_2 + 8\,OH^- + \cancel{6\,e^-}$$

The results are

$$2\,MnO_4^- + 6\,I^- + 4\,H_2O \longrightarrow 2\,MnO_2 + 3\,I_2 + 8\,OH^-$$

9. Check the equation to see that it is balanced.

	Left	Right
Mn	2	2
I	6	6
O	12	12
H	8	8
Charge	-8	-8

The balanced equation, including phase labels, is

$$2\,MnO_4^-(aq) + 6\,I^-(aq) + 4\,H_2O(l) \longrightarrow 2\,MnO_2(s) + 3\,I_2(aq) + 8\,OH^-(aq)$$

16.8 Since lead is above copper in the activity series (Table 16.2), lead metal (Pb) will be oxidized and copper ions (Cu^{2+}) will be reduced. Thus, the lead electrode is the anode and the copper electrode is the cathode. Electrons will flow from the lead half-cell through the wire to the copper half-cell. This leads to the following half-cell reactions.

$$Pb(s) \longrightarrow Pb^{2+}(aq) + 2\ e^- \quad \text{(oxidation)}$$
$$Cu^{2+}(aq) + 2\ e^- \longrightarrow Cu(s) \quad \text{(reduction)}$$

Add the two half-reactions together and cancel two electrons from both sides of the overall reaction. The final results for the overall chemical equation is

$$Pb(s) + Cu^{2+}(aq) \longrightarrow Pb^{2+}(aq) + Cu(s)$$

■ Answers to Questions to Test Your Reading

16.1 An oxidation-reduction reaction is a reaction in which electrons are transferred from one substance to another. Oxidation is the loss of electrons while reduction is the gain of electrons. Oxidation cannot occur without reduction. Electrons can only be transferred from one species to another. Thus, one substance must lose electrons and the other gains electrons.

16.2 An oxidizing agent is an agent that causes the oxidation of another substance in an oxidation-reduction reaction. It is reduced during the process. A reducing agent is an agent that causes the reduction of another substance in an oxidation-reduction reaction. It is oxidized in the process.

16.3 The activity series is an arrangement of the elements in order of their ease of losing electrons during reactions in aqueous solutions.

16.4 Since lead is above copper in the table, lead metal will react with Cu^{2+} ions. A free element (in this case Pb) in the table will react with a monatomic ion from another element (in this case Cu^{2+}) in the table if the free element is above the other element in the activity series. Thus

$$Pb(s) + Cu^{2+}(aq) \longrightarrow Pb^{2+}(aq) + Cu(s)$$

16.5 Since tin is above hydrogen in the activity series, tin will dissolve in acid.

$$Sn(s) + 2\ H^+(aq) \longrightarrow Sn^{2+}(aq) + H_2(g)$$

16.6 Powdered aluminum would be a reducing agent, since it will lose electrons.

$$Al \longrightarrow Al^{3+} + 3\ e^-$$

Since aluminum is being oxidized, it is a reducing agent.

16.7 Elements near the top of Table 16.2 lose electrons most easily (are oxidized readily), and are the strongest reducing agents. Elements near the bottom of the table are the weakest reducing agents. Since copper is below cadmium in the activity series, copper is a weaker reducing agent than cadmium.

16.8 An oxidation number is either

    i. the charge on an atom or monatomic ion, or

    ii. the charge on an atom in a substance if the shared pair of electrons belonged to the more electronegative atom in the bond.

This concept is useful as a bookkeeping device to keep track of electrons in oxidation-reduction reactions. It is also helpful in identifying the oxidizing agent and the reducing agent. It is also sometimes used in the naming of compounds.

16.9 The oxidation number of a Cr atom in the element is zero. So are the oxidation numbers of an O atom in O_2, a P atom in P_4, and an S atom in S_8 as these are all elements.

16.10 In $MgCl_2$, the oxidation number of Mg is +2, because it is a Group IIA element (Rule 7). The oxidation number of each Cl atom is -1 because it is a Group VIIA element (Rule 4). The sum of the oxidation numbers in the compound is zero. In HI, hydrogen (Group IA) has an oxidation number of +1 and iodine (Group VIIA) has an oxidation number of -1. The sum of the oxidation numbers in the compound is zero.

16.11 In H_2O, hydrogen has an oxidation number of +1 and oxygen has an oxidation number of -2. The sum of the oxidation numbers in the compound is zero. For hydrogen peroxide, H_2O_2, hydrogen has an oxidation number of +1 and oxygen has an oxidation number of -1. The sum of the oxidation numbers in the compound is zero.

16.12 During oxidation, the oxidation number of an atom increases. During reduction, it decreases.

16.13 The reaction of hydrochloric acid with sodium hydroxide is not an oxidation-reduction reaction. It is a double displacement reaction. No oxidation numbers are changing in this reaction and no electrons are being transferred.

16.14 Coal (carbon) burning in oxygen is an oxidation-reduction reaction. The reaction can be written as

$$C(s) + O_2(g) \longrightarrow CO_2(g)$$

The oxidation numbers of carbon and oxygen in the reactants are both zero (Rule 1). In carbon dioxide, oxygen has an oxidation number of -2 and carbon is +4. Note that the oxidation number for carbon changes from 0 to +4, so it is being oxidized by oxygen. The oxidation number of oxygen changes from 0 to -2, so it is being reduced by the carbon. Therefore, oxygen is the oxidizing agent and carbon is the reducing agent.

16.15 Half-reactions are the reactions that result when an oxidation-reduction reaction has been separated into two equations, one involving the species that is oxidized (loses electrons) and one involving the species that is reduced (gains electrons). For the reaction

$$Zn(s) + Cu^{2+}(aq) \longrightarrow Zn^{2+}(aq) + Cu(s)$$

The oxidation half-reaction is

$$Zn(s) \longrightarrow Zn^{2+}(aq) + 2e^-$$

The reduction half-reaction is

$$Cu^{2+}(aq) + 2e^- \longrightarrow Cu(s)$$

Note that half-reactions involve electrons where complete reactions do not.

16.16 The two methods that are used to balance oxidation-reduction reactions are by inspection, or by the half-reaction method.

16.17 An electrochemical cell is an apparatus that either generates or uses an electric current. A voltaic cell (also called a galvanic cell) is a cell in which a spontaneous reaction generates an electric current. An electrolytic cell is a cell in which an external electric current drives a nonspontaneous oxidation-reduction reaction (electrolysis).

16.18 A spontaneous process is a physical or chemical change that occurs by itself. They do not require an outside agency (such as a source of energy) to make the reaction happen. An example from everyday life would be the spoiling of food when it is not refrigerated. Food spoils spontaneously. We put it in the refrigerator to try and slow down the process.

16.19 A half-cell is that portion of an electrochemical cell in which a half-reaction takes place. Some examples of half-reactions are

$$Zn(s) \longrightarrow Zn^{2+}(aq) + 2\,e^- \quad \text{(oxidation)}$$
$$Cu^{2+}(aq) + 2\,e^- \longrightarrow Cu(s) \quad \text{(reduction)}$$

16.20 The site of oxidation is the anode, and the site of reduction is the cathode. It is the same for both an electrolytic cell and a voltaic cell.

16.21 A battery is a single voltaic cell or a series of voltaic cells. If more than one cell is used, they are arranged so that the anode of one is connected to the cathode of another. The voltage from such a series of cells is equal to the voltage from one cell multiplied by the number of cells.

16.22 The substance that is oxidized in a lead storage battery is Pb(s). The substance being reduced is $PbO_2(s)$. When this battery is dead, the chemical reaction has reached equilibrium. When it is recharged, an external circuit drives the reaction in the reverse direction, where it is ready to be spontaneous again.

16.23 The reaction that would produce $H_2(g)$ would be

$$2\,H_2O(l) \longrightarrow 2\,H_2(g) + O_2(g)$$

16.24 In a nickel-cadmium battery, the cadmium is oxidized and the NiO(OH) is reduced.

16.25 If you fasten a wire across the terminals of a nickel-cadmium battery, the reaction

$$Cd + 2\,NiO(OH) + 2\,H_2O \longrightarrow Cd(OH)_2 + 2\,Ni(OH)_2$$

will proceed from left to right until the voltage of the cell drops to zero. Also, as a result of the reaction, heat will be generated.

16.26 In a mercury battery, the zinc is oxidized and the mercury is reduced.

■ Solutions to Practice Problems

16.27a. The charge on tin increases ($Sn^{2+} \to Sn^{4+}$) so Sn^{2+} is oxidized (lost electrons) and is the reducing agent. The charge on cerium decreases ($Ce^{4+} \to Ce^{2+}$) so Ce^{4+} is reduced (gained electrons) and is the oxidizing agent.

198 ■ CHAPTER 16

b. The charge on zinc increases (Zn → Zn^{2+}) so Zn is oxidized (lost electrons) and is the reducing agent. The charge on hydrogen decreases (2 H^+ → H_2) so H^+ is reduced (gained electrons) and is the oxidizing agent.

c. The charge on cadmium increases (Cd → Cd^{2+}) so Cd is oxidized (lost electrons) and is the reducing agent. The charge on lead decreases (Pb^{2+} → Pb) so Pb^{2+} is reduced (gained electrons) and is the oxidizing agent.

d. The charge on aluminum increases (Al → Al^{3+}) so Al is oxidized (lost electrons) and is the reducing agent. The charge on chlorine decreases (Cl_2 → 2 Cl^-) so Cl_2 is reduced (gained electrons) and is the oxidizing agent.

16.29 a. Lead is below magnesium in the activity series. Therefore, a reaction between Pb(s) and Mg^{2+}(aq) will not occur.

b. Zinc is above silver in the activity series. Therefore, a reaction between Zn(s) and Ag^+(aq) will occur.

c. Chromium is above hydrogen in the activity series. Therefore, a reaction between Cr(s) and H^+(aq) will occur.

d. Gold is below hydrogen in the activity series. Therefore, a reaction between Au(s) and H^+(aq) will not occur.

16.31

a. HBr
 +1 -1

b. Na_2O
 +1 -2

c. CH_4
 -4 +1

d. C_2H_6
 -3 +1

e. O_2
 0

f. O_3
 0

16.33

a. $NaClO_4$
 +1 +7 -2

b. $NaClO_3$
 +1 +5 -2

c. $NaClO_2$
 +1 +3 -2

d. NaClO
 +1 +1 -2

e. NaCl
 +1 -1

f. Cl_2
 0

OXIDATION-REDUCTION REACTIONS ■ 199

16.35 a. Br^- : -1

b. N^{3-} : -3

c. NH_2^- : $N = -3$, $H = +1$

d. OH^- : $O = -2$, $H = +1$

e. MnO_4^- : $Mn = +7$, $O = -2$

f. HS^- : $H = +1$, $S = -2$

16.37 a. $H_2PO_4^-$: $H = +1$, $P = +5$, $O = -2$

b. HPO_4^{2-} : $H = +1$, $P = +5$, $O = -2$

c. PO_4^{3-} : $P = +5$, $O = -2$

d. NH_4^+ : $N = -3$, $H = +1$

e. NO_3^- : $N = +5$, $O = -2$

f. NO_2^- : $N = +3$, $O = -2$

16.39a. Assign oxidation numbers:

$$N_2(g) + O_2(g) \longrightarrow 2\ NO(g)$$

$N_2: 0$; $O_2: 0$; $NO: N = +2, O = -2$

Since the oxidation number of nitrogen increases, it is being oxidized. Oxygen is being reduced. Therefore, N_2 is the reducing agent and O_2 is the oxidizing agent.

b. Assign oxidation numbers:

$$P_4(s) + 5\ O_2(g) \longrightarrow P_4O_{10}(s)$$

$P_4: 0$; $O_2: 0$; $P_4O_{10}: P = +5, O = -2$

Since the oxidation number of phosphorus increases, it is being oxidized. Oxygen is being reduced. Therefore, P_4 is the reducing agent and O_2 is the oxidizing agent.

c. Assign oxidation numbers:

$$P_4O_{10}(s) + 6\ H_2O(l) \longrightarrow 4\ H_3PO_4(aq)$$

$P_4O_{10}: P = +5, O = -2$; $H_2O: H = +1, O = -2$; $H_3PO_4: H = +1, P = +5, O = -2$

Since the oxidation numbers for phosphorus, hydrogen, and oxygen do not change during the reaction, this equation does not represent an oxidation-reduction reaction.

d. Assign oxidation numbers:

$$2\ CO(g) + O_2(g) \longrightarrow 2\ CO_2(g)$$

Under CO: +2, -2; under O_2: 0; under CO_2: +4, -2

Since the oxidation number of carbon increases, it is being oxidized. Oxygen is being reduced. Therefore, CO is the reducing agent and O_2 is the oxidizing agent.

16.41 a. $4\ Li(s) + O_2(g) \longrightarrow 2\ Li_2O(s)$

O_2 is the oxidizing agent and Li is the reducing agent.

b. $4\ Al(s) + 3\ O_2(g) \longrightarrow 2\ Al_2O_3(s)$

O_2 is the oxidizing agent and Al is the reducing agent.

c. $2\ Cr(s) + 6\ HCl(aq) \longrightarrow 2\ CrCl_3(aq) + 3\ H_2(g)$

HCl is the oxidizing agent and Cr is the reducing agent.

d. $CH_4(g) + 2\ O_2(g) \longrightarrow CO_2(g) + 2\ H_2O(l)$

O_2 is the oxidizing agent and CH_4 is the reducing agent.

16.43 a. $Br_2(aq) + SO_2(aq) + 2\ H_2O(l) \longrightarrow 2\ HBr(aq) + H_2SO_4(aq)$

Br_2 is the oxidizing agent and SO_2 is the reducing agent.

b. $6\ HI(aq) + 2\ HNO_3(aq) \longrightarrow 3\ I_2(aq) + 2\ NO(g) + 4\ H_2O(l)$

HNO_3 is the oxidizing agent and HI is the reducing agent.

c. $MnO_2(s) + 4\ HBr(aq) \longrightarrow MnBr_2(aq) + Br_2(aq) + 2\ H_2O(l)$

MnO_2 is the oxidizing agent and HBr is the reducing agent.

d. $3\ CuO(s) + 2\ NH_3(g) \longrightarrow 3\ Cu(s) + N_2(g) + 3\ H_2O(g)$

CuO is the oxidizing agent and NH_3 is the reducing agent.

16.45a. $Cr_2O_7^{2-}(aq) + 6\ Fe^{2+}(aq) + 14\ H^+(aq) \longrightarrow 2\ Cr^{3+}(aq) + 6\ Fe^{3+}(aq) + 7\ H_2O(l)$
$Cr_2O_7^{2-}$ is the oxidizing agent and Fe^{2+} is the reducing agent.

b. $2\ VO_2^+(aq) + Zn(s) + 4\ H^+(aq) \longrightarrow 2\ VO^{2+}(aq) + Zn^{2+}(aq) + 2\ H_2O(l)$
VO_2^+ is the oxidizing agent and Zn is the reducing agent.

c. $VO_2^+(aq) + Zn(s) + 4\ H^+(aq) \longrightarrow V^{3+}(aq) + Zn^{2+}(aq) + 2\ H_2O(l)$
VO_2^+ is the oxidizing agent and Zn is the reducing agent.

d. $2\ VO_2^+(aq) + 3\ Zn(s) + 8\ H^+(aq) \longrightarrow 2\ V^{2+}(aq) + 3\ Zn^{2+}(aq) + 4\ H_2O(l)$
VO_2^+ is the oxidizing agent and Zn is the reducing agent.

16.47a. $48\ NO_3^-(aq) + S_8(s) + 32\ H^+(aq) \longrightarrow 48\ NO_2(g) + 8\ SO_4^{2-}(aq) + 16\ H_2O(l)$
NO_3^- is the oxidizing agent and S_8 is the reducing agent.

b. $2\ MnO_4^-(aq) + 5\ SO_2(g) + 2\ H_2O(l) \longrightarrow 2\ Mn^{2+}(aq) + 5\ SO_4^{2-}(aq) + 4\ H^+(aq)$
MnO_4^- is the oxidizing agent and SO_2 is the reducing agent.

c. $Cr_2O_7^{2-}(aq) + 3\ HNO_2(aq) + 5\ H^+(aq) \longrightarrow 2\ Cr^{3+}(aq) + 3\ NO_3^-(aq) + 4\ H_2O(l)$
$Cr_2O_7^{2-}$ is the oxidizing agent and HNO_2 is the reducing agent.

d. $2\ MnO_4^-(aq) + 5\ HNO_2(aq) + H^+(aq) \longrightarrow 2\ Mn^{2+}(aq) + 5\ NO_3^-(aq) + 3\ H_2O(l)$
MnO_4^- is the oxidizing agent and HNO_2 is the reducing agent.

16.49a. $8\ Al(s) + 3\ NO_3^-(aq) + 5\ OH^-(aq) + 18\ H_2O(l) \longrightarrow 8\ Al(OH)_4^-(aq) + 3\ NH_3(aq)$
NO_3^- is the oxidizing agent and Al is the reducing agent.

b. $S^{2-}(aq) + 4\ I_2(aq) + 8\ OH^-(aq) \longrightarrow SO_4^{2-}(aq) + 8\ I^-(aq) + 4\ H_2O(l)$
I_2 is the oxidizing agent and S^{2-} is the reducing agent.

202 ■ CHAPTER 16

c. $Mn(OH)_2(s) + H_2O_2(aq) \longrightarrow MnO_2(s) + 2 H_2O(l)$

H_2O_2 is the oxidizing agent and $Mn(OH)_2$ is the reducing agent.

d. $Cl_2(g) + IO_3^-(aq) + 2 OH^-(aq) \longrightarrow 2 Cl^-(aq) + IO_4^-(aq)\ H_2O(l)$

Cl_2 is the oxidizing agent and IO_3^- is the reducing agent.

16.51 Since tin is above copper in the activity series, electrons will flow from the tin to the copper ions. Thus, Sn will be oxidized and Cu^{2+} will be reduced. Since oxidation occurs at the tin-tin(II) electrode, it is the anode. The copper-copper(II) electrode is the cathode since reduction occurs there. The overall reaction is

$$Sn(s) + Cu^{2+}(aq) \longrightarrow Sn^{2+}(aq) + Cu(s)$$

16.53 Chromium is above lead on the activity series. Therefore, chromium will be oxidized and lead will be reduced. The anode and cathode are

$Pb^{2+} + 2 e^- \longrightarrow Pb$ (cathode)

$Cr \longrightarrow Cr^{2+}(aq) + 2 e^-$ (anode)

The overall reaction is

$$Cr(s) + Pb^{2+}(aq) \longrightarrow Cr^{2+}(aq) + Pb(s)$$

16.55 The half-reactions are

$Al^{3+} + 3 e^- \longrightarrow Al$

$2 Br^- \longrightarrow Br_2 + 2 e^-$

The overall reaction is

$$2 AlBr_3 \longrightarrow 2 Al + 3 Br_2$$

■ Solutions to Additional Problems

16.57 $2 Cr(OH)_3(s) + 3 H_2O_2(aq) + 4 OH^-(aq) \longrightarrow 2 CrO_4^{2-}(aq) + 8 H_2O(l)$

16.59 $4 Fe(OH)_2(s) + O_2(g) + 2 H_2O(l) \longrightarrow 4 Fe(OH)_3(s)$

16.61 Follow the steps as outlined in the book.

1. Assign oxidation numbers to each atom in the equation and decide which atoms are oxidized and which are reduced.

$$Cl_2 \longrightarrow Cl^- + ClO^-$$
$$0 -1 +1 -2$$

In this equation, chlorine is oxidized and also reduced.

2. Split the skeletal equation into a half-reaction for oxidation and a half-reaction for reduction.

$$Cl_2 \longrightarrow Cl^- \quad \text{(Reduction)}$$
$$Cl_2 \longrightarrow ClO^- \quad \text{(Oxidation)}$$

3. Balance all of the atoms in each half-reaction except hydrogen and oxygen.

$$Cl_2 \longrightarrow 2\,Cl^-$$
$$Cl_2 \longrightarrow 2\,ClO^-$$

4. Balance oxygen by adding H_2O where needed.

$$Cl_2 \longrightarrow 2\,Cl^-$$
$$Cl_2 + 2\,H_2O \longrightarrow 2\,ClO^-$$

5. Balance hydrogen by adding H^+ where needed.

$$Cl_2 \longrightarrow 2\,Cl^-$$
$$Cl_2 + 2\,H_2O \longrightarrow 2\,ClO^- + 4\,H^+$$

Notice that 4 H^+ occurs in the oxidation half-reaction. Since the reaction occurs in basic solution, change the half-reactions from acidic to basic conditions by adding 4 OH^- to each side of the oxidation half-reaction.

$$Cl_2 + 2\,H_2O + 4\,OH^- \longrightarrow 2\,ClO^- + 4\,H^+ + 4\,OH^-$$

(continued)

Combine the 4 H⁺ and 4 OH⁻ on the right side to give 4 H₂O molecules. Cancel two water molecules from each side of the half-reaction.

$$Cl_2 + \cancel{2H_2O} + 4\,OH^- \longrightarrow 2\,ClO^- + \underset{2}{\cancel{4}}\,H_2O$$

The result is

$$Cl_2 + 4\,OH^- \longrightarrow 2\,ClO^- + 2\,H_2O$$

6. Balance charge by adding electrons where needed.

$$Cl_2 + 2\,e^- \longrightarrow 2\,Cl^-$$
$$Cl_2 + 4\,OH^- \longrightarrow 2\,ClO^- + 2\,H_2O + 2\,e^-$$

7. Multiply the half-reactions by a factor that will lead to the same number of electrons in each half-reaction. Since the half-reactions already have the same number of electrons (two), this step is not necessary.

8. Add the two half-reactions. Cancel 2 e⁻ on both sides of the overall equation.

$$Cl_2 + 2\,e^- \longrightarrow 2\,Cl^-$$
$$Cl_2 + 4\,OH^- \longrightarrow 2\,ClO^- + 2\,H_2O + 2\,e^-$$
$$\overline{Cl_2 + Cl_2 + 4\,OH^- + \cancel{2\,e^-} \longrightarrow 2\,Cl^- + 2\,ClO^- + 2\,H_2O + \cancel{2\,e^-}}$$

The two chlorines can be added together. The result is

$$2\,Cl_2 + 4\,OH^- \longrightarrow 2\,Cl^- + 2\,ClO^- + 2\,H_2O$$

Also, each coefficient in the equation can be divided by two. The result is

$$Cl_2 + 2\,OH^- \longrightarrow Cl^- + ClO^- + H_2O$$

9. Check the equation to make sure it is balanced.

	Left	Right
Cl	2	2
O	2	2
H	2	2
Charge	-2	-2

The balanced equation, including phase labels, is

$$Cl_2(aq) + 2\,OH^-(aq) \longrightarrow Cl^-(aq) + ClO^-(aq) + H_2O(l)$$

Solutions to Practice in Problem Analysis

16.1 The electric shock was caused by an electric current, so a spontaneous electrochemical reaction must have taken place. The alternating zinc and silver disks, along with the salt water paper disks, constitute a battery.

16.2 In order for a battery to be rechargeable, the electrochemical reaction must be reversible. For lead storage batteries and nickel-cadmium batteries, the reaction can be reversed by applying an external electric current, or a battery charger. In the case of a mercury battery, this is not possible.

Answers to the Practice Exam

1. c   2. c   3. c   4. a   5. d   6. c   7. c   8. a   9. b   10. a

11. e   12. a   13. a   14. d   15. b

17. NUCLEAR CHEMISTRY

Solutions to Exercises

17.1 The atomic numbers of potassium and calcium are 19 and 20, respectively. Their symbols are $^{40}_{19}K$ and $^{40}_{20}Ca$. The symbol for a beta particle is $^{0}_{-1}e$. The correct equation is

$$^{40}_{19}K \longrightarrow {}^{40}_{20}Ca + {}^{0}_{-1}e$$

17.2 The atomic number of plutonium is 94. The symbol for plutonium-239 is $^{239}_{94}Pu$. The symbol for an alpha particle is $^{4}_{2}He$. If we write $^{A}_{Z}X$ for the unknown nuclide, the nuclear equation is

$$^{239}_{94}Pu \longrightarrow {}^{A}_{Z}X + {}^{4}_{2}He$$

Because A and Z must be conserved, the values of A on both sides of the equation must give the same sum, as must the value of Z. Use these relationships to identify $^{A}_{Z}X$ by writing

	Left Side	Right Side	
A:	239	= A + 4	or A = 235
Z:	94	= Z + 2	or Z = 92

From the periodic table, the element with atomic number 92 is uranium. The symbol for the product is $^{235}_{92}U$, and the complete nuclear equation is

$$^{239}_{94}Pu \longrightarrow {}^{235}_{92}U + {}^{4}_{2}He$$

17.3   The atomic number of phosphorus is 15. The symbol for phosphorus-30 is $^{30}_{15}P$ and the symbol for an alpha particle is $^{4}_{2}He$. If we write $^{A}_{Z}X$ for the unknown nuclide, the nuclear equation is

$$^{A}_{Z}X + ^{4}_{2}He \longrightarrow ^{30}_{15}P + ^{1}_{0}n$$

Because A and Z must be conserved, the values of A on both sides of the equation must give the same sum, as must the value of Z. Use these relationships to identify $^{A}_{Z}X$ by writing

	Left Side	Right Side	
A:	A + 4	= 30 + 1	or A = 27
Z:	Z + 2	= 15 + 0	or Z = 13

From the periodic table, the element with atomic number 13 is aluminum. The symbol for the product is $^{27}_{13}Al$.

17.4   After one half-life, ½ of the original sample will remain. After two half-lives, ½ x ½ = ¼ of the original atoms will remain. Following this pattern, after four half-lives, ½ x ½ x ½ x ½ = 1/16 of the original sample will be left. Therefore, the number of atoms remaining after four half-lives will be

1/16 x 32,000,000 atoms = 2,000,000 atoms

17.5   We must first determine how many half-lives have passed. If the original number of disintegrations per minute is 16, then after one half-life it will be reduced to 8 disintegrations per minute, and after two half-lives, 4 disintegrations per minute. Thus, two half-lives must have passed. Since each half-life is approximately 5730 years, the total amount of time that has passed since the animal died is

2 x 5730 years = 11,460 = 11,400 years

Thus, the animal must have died in

2000 - 11,400 = -9400 AD   or   9400 BC

■ Answers to Questions to Test Your Reading

17.1   Radioactivity is the emission of radiation in the form of particles or energy coming from the nucleus of an atom undergoing spontaneous disintegration.

17.2 The atomic number of nitrogen-14 (and all other forms of nitrogen) is 7. Its mass number is 14. The number of protons in the nucleus is equal to the atomic number, 7. The number of neutrons is 14 - 7 or 7. The number of nucleons is equal to the mass number, 14. The correct symbol is $^{14}_{7}N$.

17.3 Isotopes are atoms whose nuclei have the same atomic number but different mass number. A nuclide is an nucleus of a specific isotope.

17.4 Radioactive decay is the spontaneous disintegration of a nucleus. The four types of radiation that are emitted during radioactive decay are alpha emission, beta emission, positron emission, and gamma emission.

17.5 A radioactive decay series is a sequence of decay steps that continues until a stable nucleus is reached.

17.6 The three naturally occurring radioactive series begin with uranium-238, uranium-235, and thorium-232.

17.7 The last member of the uranium-238 radioactive decay series is lead-206. It is formed in two different processes. First, thallium-206 decays by emitting a beta particle.

$$^{206}_{81}Tl \longrightarrow\ ^{206}_{82}Pb + ^{0}_{-1}e$$

The second process involves polonium-210. The nuclide decays by emitting an alpha particle.

$$^{210}_{84}Po \longrightarrow\ ^{206}_{82}Pb + ^{4}_{2}He$$

Both processes produce lead-206 as a final, stable product.

17.8 Two naturally occurring nuclides that are not in any of the naturally occurring radioactive decay series are carbon-14 and potassium-40.

17.9 A particle accelerator is a device used to accelerate electrons, protons, and alpha particles to very high speeds. They are used in certain nuclear reactions to impart enough energy to the particle so that it can penetrate the nucleus of an element with a large atomic number.

17.10 The discovery of the transuranium elements has allowed scientists to extend the periodic chart up to element 109. All of the new elements are radioactive. Perhaps the most important of these elements is plutonium, atomic number 94, which is at the heart of the atomic weapons program. Also, as the seventh row (period) of the table is completed, it is hoped that new, stable elements might be discovered. These will be called the superheavy elements. Element with atomic number 118 will fall into the Noble gas Group (VIII).

17.11 A half-life is the time required for half of the original sample of nuclei to decay. Thus, for cesium-137, with half-life of 30.2 years, after this length of time, one half of the original sample will have decayed away.

17.12 Technetium must be prepared in the laboratory because no isotopes of this element are members of a naturally occurring radioactive decay series. Uranium-238 is found on earth because its half-life, 4.51×10^9 years, is approximately the same as the age of the earth, so only one half-life for this isotope has elapsed. This means that approximately ½ of the uranium-238 that was present when the earth was formed is still on the earth. The longest lived isotope of technetium is much smaller (around 1 million years) so that any of this isotope that was present when the earth formed has long since decayed away and there is none left.

17.13 If an equal number of atoms of each nuclide in Table 17.2 are initially present, then after one billion years had elapsed, the nuclide that would be present in the largest amount would be the one with the longest half-life, or uranium-238. The nuclide that would be present in the smallest amount would be the one with the shortest half-life, or polonium-214.

17.14 Dating is a technique for determining the age of certain old objects that relies on the radioactive nuclides in the object. The number of half-lives that have elapsed since the object was formed multiplied by the value of one half-life is the age of the object.

17.15 Since one-half as much carbon-14 as the uncut tree is present, one half-life has elapsed. The age of the wood is the same as the half-life of carbon-14, or 5730 years.

17.16 If no detectable carbon-14 is present, the wood is at least 50,000 years old. The actual age of the wood cannot be determined precisely.

17.17 A Geiger-Müller counter consists of a metal tube, which contains argon, fitted with a thin window through which radiation enters. A positively charged wire runs down the center of the tube, and the wall of the tube is negatively charged. When alpha particles, beta particles, or gamma rays enters the tube, it creates an electrical current that causes a clicking in the counter. A scintillation counter is a device for detecting alpha particles, beta particles, or gamma rays. It contains a substance such as sodium iodide or zinc sulfide that emits a flash of light each time it is struck with radiation. Each flash is detected and counted.

17.18 If radioactive phosphorus were administered to a living organism, it could be used as a radioactive tracer. Any radioactive nuclide can function as a tracer but it is desirable to have a nuclide with a relatively short half-life of a few days so that the organism is not subjected to radiation for a long period of time. Under these conditions, the phosphorus can be used to trace the pathway of the substance in the body. You could obtain information on where in the body the phosphorus is absorbed, how long it stays in the body, and the pathology of those parts of the body that absorb the phosphorus and use it as a nutrient.

17.19 The principal sources of radiation that the average person is exposed to include radon in the air, uranium and other radioactive elements in soil and rocks, and cosmic rays. Also, there is potassium-40 and carbon-14 in the body. About 55% of the average annual radiation comes from radon in the air.

17.20 A rem (from roentgen equivalent for man) is a unit for a radiation dose. Doses of 25 to 200 rems can cause a decrease in the number of white blood cells and doses up to 500 rems or more can cause death. Since the average annual exposure of a person in the United States to radiation from natural and human sources is about 0.360 rems, exposure to 0.10 rems of radioactivity would not have serious implications to you. According to the Nuclear Regulatory Commission, exposure to 0.360 rems corresponds to an annual risk of death from radiation-induced cancer of about 1800 in 10 million. Thus, exposure to 0.10 rems of radiation would have much less risk. However, I would still ask questions like "What type of radiation was I exposed to?", "Are there any short term effects to exposure to low doses of radiation?", "Are there any cumulative effects to exposure to low doses of radiation?", " Can the body repair the damage done by exposure to low doses of radiation?", "How confident is the Nuclear Regulatory Commission about its report of risk?", etc.

17.21 Radon itself is unreactive and can be inhaled without any chemical effects. However, radon-222 decays by emission of an alpha particle, which can be very dangerous when they are inside the body. Also, the product of the decay is polonium-218, which is a solid that will then be deposited in the lungs. It decays by emitting an alpha particle. Many scientists believe that inhaling radon can cause lung cancer.

17.22 Nuclear fission is a process in which a nucleus with mass number greater than 50 splits to form two lighter and more stable nuclei. Nuclear fusion is a process in which light nuclei, with mass numbers less than 50, combine to give heavier, more stable nuclei. Energy is released during both processes.

17.23 Since carbon-12 is a nuclei with mass number less than 50, we would expect it to undergo fusion with other carbon-12 nuclei, rather than fission into smaller nuclei.

17.24 Fission of uranium-235 is induced by the absorption of neutrons, often from an outside source. As the nuclide splits by its many pathways, more neutrons are produced, which then are absorbed by other uranium-235 nuclides causing a chain reaction.

17.25 Critical mass is the minimum mass that will allow fission to become a chain reaction. If the mass is larger than the critical mass, the chain reaction will continue rapidly and become supercritical and explode.

17.26 A nuclear power plant is not a potential atomic bomb waiting to explode. The fuel used is only enriched to about 3% uranium-235. This is not enough nuclides to reach critical mass and an uncontrolled chain reaction is not possible.

Solutions to Practice Problems

17.27 $^{87}_{37}Rb \longrightarrow \, ^{87}_{38}Sr + \, ^{0}_{-1}e$

17.29 $^{232}_{90}Th \longrightarrow \, ^{228}_{88}Ra + \, ^{4}_{2}He$

17.31 If we let $^{A}_{Z}X$ represent the unknown product nuclide, we can write the decay process as

$$^{210}_{84}Po \longrightarrow \, ^{A}_{Z}X + \, ^{4}_{2}He$$

Since A and Z must be conserved we can write

A: $210 = A + 4$, or $A = 206$
Z: $84 = Z + 2$, or $Z = 82$

Therefore, the product nuclide must be lead-206. The nuclear equation can now be written as

$$^{210}_{84}Po \longrightarrow \, ^{206}_{82}Pb + \, ^{4}_{2}He$$

17.33 If we let $^{A}_{Z}X$ represent the unknown product radiation, we can write the decay process as

$$^{18}_{9}F \longrightarrow \, ^{18}_{8}O + \, ^{A}_{Z}X$$

Since A and Z must be conserved, we can write

A: $18 = 18 + A$, or $A = 0$
Z: $9 = 8 + Z$, or $Z = 1$

The product radiation must have a mass number of zero and a charge of +1. This is a positron. Thus we can write the nuclear equation as

$$^{18}_{9}F \longrightarrow \, ^{18}_{8}O + \, ^{0}_{1}e$$

17.35 If we let ^A_ZX represent the unknown product nuclide, we can write the nuclear equation as

$$^6_3Li + ^1_0n \longrightarrow ^3_1H + ^A_ZX$$

Since A and Z must be conserved, we can write

A: 6 + 1 = 3 + A,   or A = 4
Z: 3 + 0 = 1 + Z,   or Z = 2

Thus, the unknown nuclide must be ^4_2He, an alpha particle.

17.37 If we let ^A_ZX represent the unknown target nuclide, we can write the nuclear equation as

$$^A_ZX + ^4_2He \longrightarrow ^{242}_{96}Cm + ^1_0n$$

Since A and Z must be conserved, we can write

A: A + 4 = 242 + 1,   or A = 239
Z: Z + 2 = 96 + 0,   or Z = 94

Thus, the unknown nuclide must be $^{239}_{94}Pu$.

17.39 After three half-lives for any nuclide, there will be ½ x ½ x ½ = ⅛ of the original sample remaining.

17.41 First, we must determine the approximate number of half-lives that have elapsed. Two weeks is 8 days, so we can write

$$8 \text{ days} \times \frac{1 \text{ half-life}}{2.69 \text{ days}} = 2.97 \text{ half-lives}$$

If we round this off to approximately 3 half-lives, then ½ x ½ x ½ = ⅛ of the original sample will remain, or

$$\frac{1}{8} \times 0.00100 \text{ g} = \underline{1}.2 \times 10^{-4} = 1 \times 10^{-4} \text{ g}$$

17.43 First, we must determine the number of half-lives that have elapsed. The disintegrations per minute will drop from 16 to 8 in one half-life. Thus, the cypress beam must be approximately one half-life, or 5730 years, old. The tree was felled approximately

2000 - 5730 = -3670 = -3700 AD   or   3700 BC

17.45 First, we must determine the approximate number of half-lives that have elapsed. The disintegrations per minute will drop from 16 to 4 in two half-lives. Thus, the artifact is approximately two half-lives, or 2 x 5730 years = 11,460 years. This result should be rounded off to 11,400 years old.

■ Solutions to Additional Problems

17.47 If we let A be the unknown mass number of the astatine nucleus, we can write

$$^{209}_{83}Bi + ^{4}_{2}He \longrightarrow ^{A}_{85}At + 2\,^{1}_{0}n$$

Since A must be conserved, we can write

A: 209 + 4 = A + 2 x 1,   or A = 211

Thus, the product nuclide must be astatine-211. The nuclear equation for this reaction is

$$^{209}_{83}Bi + ^{4}_{2}He \longrightarrow ^{211}_{85}At + 2\,^{1}_{0}n$$

17.49 If we let $^{A}_{Z}X$ represent the unknown nuclide, then to conserve A and Z, we can write

A: 238 + 12 = A + 4 x 1,   or A = 246
Z: 92 + 6 = Z + 4 x 0,   or Z = 98

The unknown nuclide must be californium-246. The nuclear equation for this reaction is

$$^{238}_{92}U + ^{12}_{6}C \longrightarrow ^{246}_{98}Cf + 4\,^{1}_{0}n$$

Solutions to Practice in Problem Analysis

17.1 In order to determine the age of the meteorite, you first must determine how many half-lives have elapsed. Uranium-238 decays by a series of steps (Figure 17.2), so it is necessary to examine the half-lives of each of the nuclides in the series (Table 17.2). According to the table, all of the half-lives of the nuclides are small compared to the half-life of uranium-238 (4.51×10^9 years). The next longest half-life is for uranium-234 (2.47×10^5 years). Therefore, you can assume that nearly all of the uranium-238 that has decayed is now in the form of lead-206. Very small amounts of the other substances will be present, but not enough to affect the result. Since approximately equal amounts of uranium-238 and lead-206 are present in the meteorite, one half-life has elapsed and the age of the meteorite is approximately 4.51×10^9 years old. This result should be rounded off.

17.2 This is a problem involving stoichiometry and gases. Since one half-life has elapsed, ½ x 1.00 g, or 0.50 g of polonium-210 has decayed. Each polonium-210 nucleus that decays produces an alpha particle, or equivalently, a helium atom. Convert this quantity to moles by dividing by the molar mass of polonium-210 (210 g/mol), and then multiply by the appropriate mole ratio (1 mol He/ 1 mol Po). The result is the number of moles of helium produced. The conditions of temperature and pressure (0°C, 1 atm) are STP, so 1 mole of gas occupies 22.4 liters. Multiply the number of moles by this conversion factor to get the volume of helium produced.

Answers to the Practice Exam

1. a   2. b   3. b   4. d   5. b   6. e   7. b   8. d   9. d   10. d
11. e   12. b   13. b   14. d   15. c

18. ORGANIC CHEMISTRY

■ Solutions to Exercises

18.1 a. The longest carbon chain is five atoms, so the parent name is pentane. There are two methyl groups attached to the parent chain.

$$CH_3\ ^3CH\ ^2CH\ ^1CH_3$$
$$\qquad\quad |\qquad |$$
$$\qquad\ ^4CH_2\ CH_3\quad CH_3$$

(with CH$_3$ branch on C2 and ethyl continuation numbered 4,5)

The name is 2,3-dimethylpentane.

b. The longest carbon chain is four atoms, so the parent name is butane. There are two methyl groups attached.

$$^4CH_3\ ^3CH\ ^2CH\ ^1CH_3$$
with CH$_3$ branches on C2 and C3.

The name is 2,3-dimethylbutane.

18.2 3-methylpentane has a parent chain of five carbons and one methyl branch on carbon 3. Fill in hydrogens to satisfy four bonds for each carbon. The final structure is

$$\begin{array}{c}\qquad\qquad CH_3\\ H\ H\ \ |\ \ H\ H\\ |\ \ |\ \ |\ \ |\ \ |\\ H-C-C-C-C-C-H\\ |\ \ |\ \ |\ \ |\ \ |\\ H\ H\ H\ H\ H\end{array}$$

18.3 a. Propane is $CH_3CH_2CH_3$; The molecular formula is C_3H_8. The balanced equation is

$$C_3H_8(g) + 5\,O_2(g) \longrightarrow 3\,CO_2(g) + 4\,H_2O(g)$$

b. There are many possibilities for the reaction of chlorine with butane, C_4H_{10}. Some are

$$CH_3CH_2CH_2CH_3 + Cl_2 \longrightarrow CH_3CH_2CH_2CH_2Cl + HCl$$
$$CH_3CH_2CH_2CH_3 + Cl_2 \longrightarrow CH_3CH_2CHClCH_3 + HCl$$

18.4 a. The longest chain is six carbons, and it has a triple bond.

$$^6CH_3\,^5CH_2\,^4CH_2\,^3C\equiv{}^2C\,^1CH_3$$

The name of the compound is 2-hexyne.

b. The longest chain is seven carbons. Number from the end nearest the double bond.

$$\begin{array}{c}CH_3\\ |\\ {}^1CH_3\,{}^2CH\,{}^3CH={}^4CH\,{}^5CH_2\,{}^6CH_2\,{}^7CH_3\end{array}$$

The name of the compound is 2-methyl-3-heptene.

18.5   a. m-chloronitrobenzene   b. 1,2,4-tribromobenzene

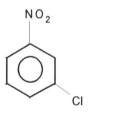
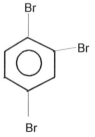

ORGANIC CHEMISTRY ■ 217

c. p-bromotoluene

d. nitrobenzene

18.6 The longest chain containing the OH group is six so the parent name is hexanol. There are two methyl groups. Begin numbering from the end closest to the OH group.

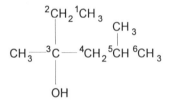

The name of the compound is 3,5-dimethyl-3-hexanol.

18.7 a. The compound is an aldehyde; the numbering of the parent chain is

$$\underset{H}{\overset{O}{\|}}\,^1C\,^2CH_2\,^3CH_2\,^4CH\,\underset{CH_3}{\overset{^5CH_2\,^6CH_3}{|}}$$

The name of the compound is 4-methylhexanal.

b. The compound is a ketone; The numbering of the parent chain is

$$^5CH_3\,^4CH_2\,^3CH_2\,^2CO\,^1CH_3$$

The name of the compound is 2-pentanone.

■ Answers to Questions to Test Your Reading

18.1 Organic chemicals around the house may include aspirin, rubbing alcohol, nail polish remover, acetone, clothing made of polyester or nylon, etc.

18.2 The present definition of organic chemistry is the chemistry of carbon compounds. The original definition of organic compounds were substances obtained from plant and animal materials.

18.3 Carbon can bond in the following four ways

methane, CH_4 formaldehyde, H_2CO carbon dioxide, CO_2 acetylene, C_2H_2

18.4 Isomers are compounds that have the same molecular formula but different structural formulas. An example would be C_2H_6O, where the two structures for ethanol and dimethyl ether are isomers.

18.5 The condensed structural formula of butanol would be $CH_3CH_2CH_2CH_2OH$.

18.6 The first four members of the alkane series are methane, CH_4, ethane, C_2H_6, propane, C_3H_8, and butane, C_4H_{10}.

18.7 A straight-chain alkane of seven carbons would be heptane, C_7H_{16}.

18.8 The alkyl group obtained from propane is propyl. (see Table 18.2)

18.9 The major chemical sources of alkanes are natural gas and petroleum. These materials occur naturally as the end products of the decay of organic material in the earth.

The major uses of natural gas and petroleum alkanes are as starting materials for the production of organic chemicals. Natural gas is used for home and industrial heating. Propane and butane are gases that are available as liquids in fuel cylinders. Gasoline is the principal product of petroleum. Other uses of petroleum include various fuels, solvents, oils and lubricants. (see Table 18.3)

18.10 Petroleum refining is a chemical process for petroleum to produce a fuel suitable for modern automobiles.

ORGANIC CHEMISTRY ■ 219

18.11 Straight-chain alkanes are unsuitable as gasoline because they tend to preignite causing the engine to "knock", which causes undue wear on an engine. They have low octane ratings.

18.12 The combustion of methane is given by the reaction

$$CH_4(g) + 2\,O_2(g) \longrightarrow CO_2(g) + 2\,H_2O(g)$$

The reaction of methane with chlorine is given by

$$CH_4 + Cl_2 \longrightarrow CH_3Cl + HCl$$

18.13 Two alkenes are ethylene, C_2H_4, and propylene, C_3H_6. Their structures are

and

An example of an alkyne is acetylene, C_2H_2, whose structure is

$$H\text{—}C\equiv C\text{—}H$$

18.14 Ethylene is present in small quantities in natural gas and petroleum. In the chemical industry it is produced from ethane by heating. It is used to prepare plastics like polyethylene.

18.15 The reaction for the addition of bromine to ethylene is

$$CH_2\!=\!CH_2 + Br_2 \longrightarrow \underset{\underset{Br\ \ Br}{|\ \ |}}{CH_2CH_2}$$

This reaction can be used as a test for multiple bonds because the red-brown color of bromine disappears as it reacts with the multiple bond.

18.16 A polymer is a very large molecule consisting of many repeating units of low molecular weight. Polyethylene is an example. It is a chain of repeating C_2H_4 units. The monomer is a compound used to prepare a polymer and gives rise to the repeating unit. Polyethylene means "many ethylene monomer units".

18.17 Polypropylene looks like

```
    CH₃ H   CH₃ H   CH₃ H
     |  |    |  |    |  |
  —  C— C — C— C  — C— C —
     |  |    |  |    |  |
     H  H    H  H    H  H
```

18.18 The resonance description of benzene is

[Two Kekulé resonance structures of benzene shown connected by a double-headed arrow]

The meaning of the double arrow is that the extra pairs of electrons represented by double bonds are actually delocalized over the benzene ring, rather than localized between two carbon atoms as shown by any single formula. All six carbon atoms share the six electrons of the three double bonds, giving the benzene molecule great stability.

18.19 Para is a prefix used in a disubstituted benzene ring where the substituents are on the carbon atoms on opposite sides of the ring.

18.20 A functional group is a reactive position of a molecule that undergoes predictable reactions. Some examples are alkenes (double bonds), alcohols (ROH), ethers (ROR'), etc. (see Table 18.4)

18.21 An alcohol is a compound ROH that contains an OH group bonded to a tetrahedral carbon atom (an alkyl group R). Some examples are methanol, CH_3OH, and ethanol CH_3CH_2OH.

ORGANIC CHEMISTRY ■ 221

18.22 Methanol is produced from the reaction of carbon monoxide and hydrogen, using a metal oxide catalyst.

$$CO(g) + 2H_2(g) \xrightarrow[400°C, 150\ atm]{ZnO\ -\ Cr_2O_3} CH_3OH(g)$$

Methanol is a common solvent for organic materials, such as shellac and varnish. It is the main ingredient in automobile windshield washer compounds, and as a fuel.

18.23 Ethanol is prepared from ethylene by reacting the ethylene with steam in the presence of an acid catalyst (such as sulfuric acid).

$$CH_2\!\!=\!\!CH_2 + HOH \xrightarrow[heat]{H^+} CH_3CH_2OH$$

18.24 1-propanol will be oxidized by an acidic dichromate ion solution into propanoic acid as a final product.

$$\begin{array}{c} H\ \ H\ \ O \\ |\ \ \ |\ \ \ \| \\ H-C-C-C-O-H \\ |\ \ \ | \\ H\ \ H \end{array}$$

18.25 a. 3-methyl-3-heptanol is a tertiary alcohol and is unreactive towards acidic dichromate ion solutions.

b. 2-methyl-3-heptanol is a secondary alcohol and will react to form a ketone.

18.26 a. aldehyde   **b.** alcohol   **c.** ketone

18.27 Formaldehyde, HCHO, is a gas with a characteristic irritating odor. An aqueous solution containing methanol is called formalin and is used as a preservative and disinfectant. A commercial use is in the manufacture of polymers.

Acetone, CH_3COCH_3, is a flammable, volatile liquid with a slight aromatic odor. It is a solvent for certain resins and acetate plastics, and is found in nail polish and nail polish remover. In industry it is used in the manufacture of methyl methacrylate, which is made into polymers like Lucite and Plexiglas.

18.28 You do not need a number for propanone, because the carbonyl group of the ketone can only be on the number two carbon atom; otherwise, it would be an aldehyde.

18.29 Acetic acid and its corresponding ionization reaction is given by

$$CH_3COOH + H_2O \rightleftharpoons CH_3COO^- + H_3O^+$$

The (—COOH) group is the carboxylic acid group.

18.30 Acetic acid + methanol ⟶ methyl acetate + water

$$CH_3COOH + HOCH_3 \xrightarrow{H^+} CH_3COOCH_3 + H_2O$$

18.31 Aniline + hydronium ion ⟶ phenylammonium ion + water

$$C_6H_5NH_2 + H_3O^+ \longrightarrow [C_6H_5NH_3]^+ + H_2O$$

18.32 Acetic acid + methylamine ⟶ methylacetamide + water

$$CH_3COOH + CH_3NH_2 \xrightarrow{\Delta} CH_3CONHCH_3 + H_2O$$

ORGANIC CHEMISTRY ■ 223

■ Solutions to Practice Problems

18.33 a. Chloroform has four bonding groups around the central atom and is tetrahedral.

b. Formaldehyde has three bonding groups around the central atom and is trigonal planar.

c. Hydrogen cyanide has two bonding groups around the central atom and is linear.

18.35 a.
```
    H  H  H  H
    |  |  |  |
H – C– C– C– C– H
    |  |  |  |
    H  H  H  H
```

```
         H
         |
      H– C– H
   H     |     H
   |     |     |
H– C     C     C– H
   |     |     |
   H     H     H
```

b.
```
    H  H  H
    |  |  |
H – C– C– C– O– H
    |  |  |
    H  H  H
```

```
         O– H
   H     |     H
   |     |     |
H– C     C     C– H
   |     |     |
   H     H     H
```

```
   H     H        H
   |     |        |
H– C     C – O – C– H
   |     |        |
   H     H        H
```

18.37 a.
```
   H  Cl
   |  |
H– C– C– Cl
   |  |
   H  H
```

```
   Cl Cl
   |  |
H– C– C– H
   |  |
   H  H
```

b.
```
   H  H  H
   |  |  |
H– C– C– C– Cl
   |  |  |
   H  H  H
```

```
   H  Cl H
   |  |  |
H– C– C– C– H
   |  |  |
   H  H  H
```

224 ■ CHAPTER 18

18.39 a.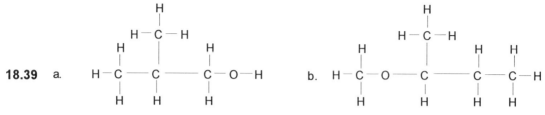

18.41 The molecular formula is C_8H_{18}. The condensed structural formula is

$CH_3CH_2CH_2CH_2CH_2CH_2CH_2CH_3$

The full structural formula is

```
    H  H  H  H  H  H  H  H
    |  |  |  |  |  |  |  |
H — C— C— C— C— C— C— C— C —H
    |  |  |  |  |  |  |  |
    H  H  H  H  H  H  H  H
```

18.43 a. 3-methylhexane   b. 2,2-dimethylbutane   c. 3-ethylhexane

18.45 a.
$$\begin{array}{c} CH_3 \\ | \\ CH_3\,CH\,CH_2CH_3 \end{array}$$

b.
$$\begin{array}{c} CH_3 \\ | \\ CH_3CHCHCH_2CH_3 \\ | \\ CH_2CH_3 \end{array}$$

18.47 $C_7H_{16} + 11\,O_2 \longrightarrow 7\,CO_2 + 8\,H_2O$

18.49 $CH_3CH_2CH_2CH_2CH_3 + Cl_2 \longrightarrow CH_3CH_2CH_2CH_2CH_2Cl + HCl$

18.51 a. 4-methyl-1-pentene   b. 3-methylbutyne

18.53 $CH_3CH{=}CH_2 + Br_2 \longrightarrow$ $\underset{}{CH_3}\overset{Br}{\underset{|}{C}}HCH_2Br$

18.55

$$\underset{F}{\overset{F}{C}}=\underset{F}{\overset{F}{C}} + \underset{F}{\overset{F}{C}}=\underset{F}{\overset{F}{C}} + \underset{F}{\overset{F}{C}}=\underset{F}{\overset{F}{C}} \longrightarrow -\underset{F}{\overset{F}{C}}-\underset{F}{\overset{F}{C}}-\underset{F}{\overset{F}{C}}-\underset{F}{\overset{F}{C}}-\underset{F}{\overset{F}{C}}-\underset{F}{\overset{F}{C}}-$$

18.57 (two resonance structures of toluene)

18.59
a. m-diethylbenzene   b. o-bromotoluene   c. 1,3,5-tribromobenzene

18.61 a. m-dibromobenzene   b. 1,2,3-tribromobenzene

18.63 a. tertiary   b. primary   c. secondary

18.65   3-methyl-2-pentanol

18.67 a. $CH_3CH_2CH_2CH_2OH + 6\,O_2 \longrightarrow 4\,CO_2 + 5\,H_2O$

b. $3\,CH_3CH_2CH_2CH_2OH + Cr_2O_7^{2-} + 8\,H^+ \longrightarrow$

$$3\,CH_3CH_2CH_2\overset{\overset{O}{\|}}{C}-H + 2\,Cr^{3+} + 7\,H_2O$$

c. 3 CH₃CH(OH)CH₂CH₂CH₃ + Cr₂O₇²⁻ + 8 H⁺ ⟶

3 CH₃C(O)CH₂CH₂CH₃ + 2 Cr³⁺ + 7 H₂O

d. no reaction - tertiary alcohols do not react with acidic dichromate solutions.

18.69 a. 2-methylbutanal   b. 3-methyl-2-butanone

18.71 a. CH₃CH₂CH(CH₃)CH(CH₃)—C(=O)—H   b. CH₃C(=O)—C(CH₃)(CH₃)—CH₂CH₃

18.73 2,2-dimethylpropanoic acid

18.75 CH₃CH(CH₃)—C(=O)—OH

18.77 CH₃CH₂CH₂C(=O)—OH + CH₃CH(OH)CH₃ —H⁺→ CH₃CH₂CH₂C(=O)—O—CH(CH₃)CH₃ + H₂O

18.79 CH₃CH₂CH₂CH₂NH₂ + H₃O⁺ ⟶ [CH₃CH₂CH₂CH₂NH₃]⁺ + H₂O

18.81 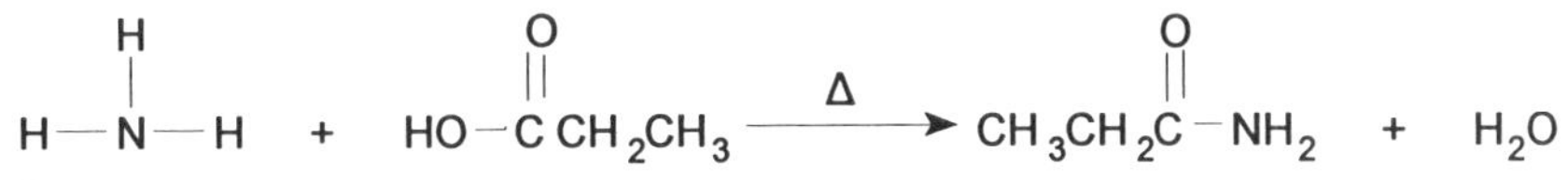

■ Solutions to Additional Problems

18.83 a. 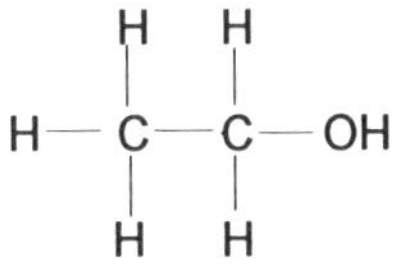

b. 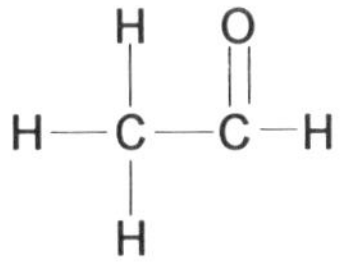

c. 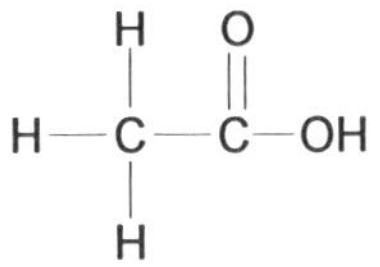

d. 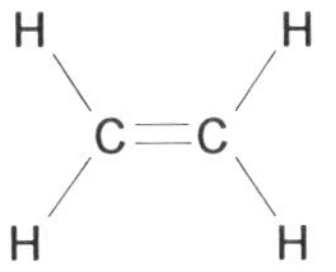

18.85 a. 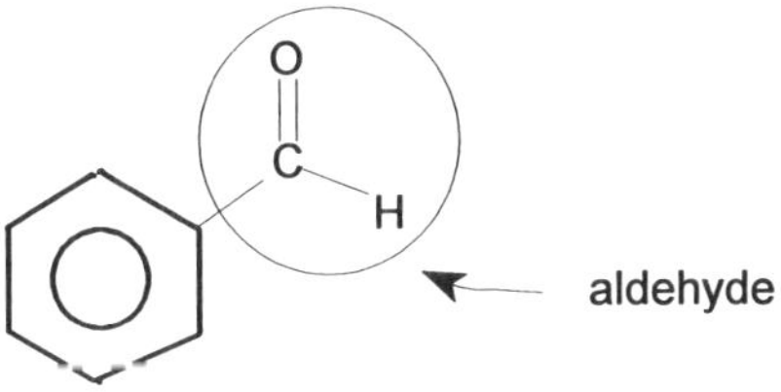
aldehyde

b. 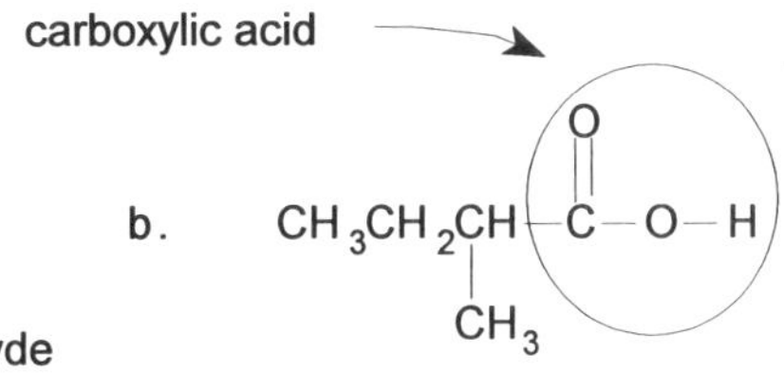
carboxylic acid

c. 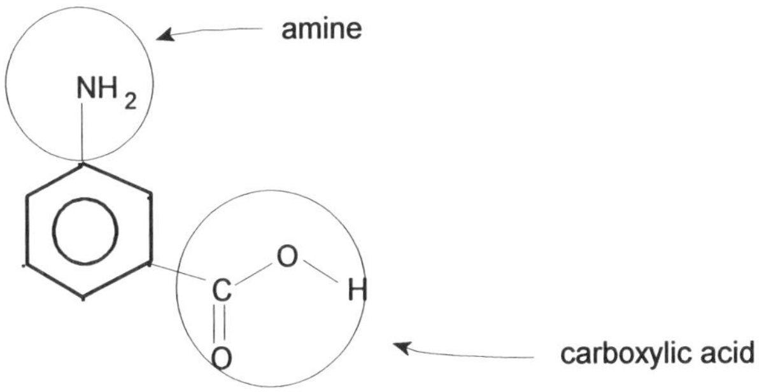
amine
carboxylic acid

18.87 a. 2,3-dimethyl-2-butene b. 3,5-dimethyloctane c. 1,3,5-triethylbenzene

18.89

a.
```
        CH₃  CH₂CH₃
         |    |
CH₃C—CH₂CHCH₂CH₂CH₂CH₃
  |
  CH₃
```

b.
```
   CH₃ CH₃
    |   |
CH₃C=CCH₂CH₂CH₂CH₃
```

c. benzene ring with CH₂CH₃, CH₂CH₃, CH₂CH₃ substituents (1,2,3-triethylbenzene)

18.91 CH₃CH₂CH₂CH₂CH₂Cl   CH₃CH₂CH₂CHCH₃   CH₃CH₂CHCH₂CH₃
                                          | |
                                          Cl   Cl

18.93 CH₃CH=CHCH₃ + Br₂ ⟶ CH₃CHCHCH₃
                                      | |
                                      Br Br

18.95
$$3\ CH_3CH_2\underset{\underset{CH_3}{|}}{\overset{\overset{CH_3}{|}}{C}}CH_2OH + 2\ Cr_2O_7^{2-} + 16\ H^+ \longrightarrow$$

$$3\ CH_3CH_2\underset{\underset{CH_3}{|}}{\overset{\overset{CH_3}{|}}{C}}\overset{O}{\overset{\|}{C}}-OH + 4\ Cr^{3+} + 11\ H_2O$$

18.97 a. ⁴CH₃³C(CH₃)(CH₃)²CH=¹CH₂   3,3-dimethyl-1-butene

b. ⁵CH₃⁴CH(CH₃)³CH₂²C(=O)¹CH₃   4-methyl-2-pentanone

18.99

$$HO-\underset{\underset{O}{\|}}{C}-(CH_2)_2-\underset{\underset{O}{\|}}{C}-OH \;+\; H-\underset{\underset{H}{|}}{N}-(CH_2)_4-\underset{\underset{H}{|}}{N}-H \;\longrightarrow$$

$$-\underset{\underset{O}{\|}}{C}-(CH_2)_2-\underset{\underset{O}{\|}}{C}-\underset{\underset{H}{|}}{N}-(CH_2)_4-\underset{\underset{H}{|}}{N}-\underset{\underset{O}{\|}}{C}-(CH_2)_2-\underset{\underset{O}{\|}}{C}-\underset{\underset{H}{|}}{N}-(CH_2)_4-\underset{\underset{H}{|}}{N}-$$

■ Solutions to Practice in Problem Analysis

18.1 The IUPAC name of this compound is obtained in the following manner. First, note that the compound consists of only carbon and hydrogen atoms, and all of the bonds are single bonds. This makes the compound an alkane. The compounds formula is C_7H_{16}, which fits the pattern for an alkane, C_nH_{2n+2}. To name the alkane, first determine the parent chain. Here, the longest carbon-atom chain is five carbons long. Next, determine what alkyl groups are present as branches (Table 18.2). This compound has an ethyl group, C_2H_5, as a branch off the parent chain. The ethyl group is located on the third carbon from the end.

18.2 To write the structural formula for 2,3-dimethylpentane, an alkane, split the name into the parent name and the names of the branch chains. For this compound the parent chain is pentane, and there are two methyl branches located on carbons 2 and 3. Add hydrogen atoms to the carbons in the parent chain to satisfy the tetravalence of carbon. Carbon atoms 1 and 3 need three hydrogen atoms each, giving CH_3. Carbon atoms 2 and 3 require one additional hydrogen atom each, giving CH. Carbon atom 4 needs two additional hydrogen atoms, giving CH_2.

■ Answers to the Practice Exam

1. c   2. b   3. b   4. e   5. d   6. b   7. c   8. e   9. d   10. a

19. BIOCHEMISTRY

■ Solutions to Exercises

19.1 Glycine and phenylalanine can react to form two possible structures. The structural formulas are

glycine + phenylalanine

$$\text{H}-\underset{\underset{\text{H}}{|}}{\overset{\overset{\text{H}}{|}}{\text{N}}}-\underset{\underset{\text{H}}{|}}{\overset{\overset{\text{H}}{|}}{\text{C}}}-\overset{\overset{\text{O}}{\|}}{\text{C}}-\underset{\underset{\text{H}}{|}}{\text{N}}-\underset{\underset{\text{CH}_2\text{-C}_6\text{H}_5}{|}}{\overset{\overset{\text{H}}{|}}{\text{C}}}-\overset{\overset{\text{O}}{\|}}{\text{C}}-\text{OH}$$

phenylalanine + glycine

$$\text{H}-\underset{\underset{\text{H}}{|}}{\overset{\overset{\text{H}}{|}}{\text{N}}}-\underset{\underset{\text{CH}_2\text{-C}_6\text{H}_5}{|}}{\overset{\overset{\text{H}}{|}}{\text{C}}}-\overset{\overset{\text{O}}{\|}}{\text{C}}-\underset{\underset{\text{H}}{|}}{\text{N}}-\underset{\underset{\text{H}}{|}}{\overset{\overset{\text{H}}{|}}{\text{C}}}-\overset{\overset{\text{O}}{\|}}{\text{C}}-\text{OH}$$

19.2

aldose

```
        O
        ‖
        C—H
        |
    H—C—OH
        |
    H—C—OH
        |
    H—C—OH
        |
    H—C—OH
        |
    H—C—OH
        |
    H—C—OH
        |
        H
```

ketose

```
        H
        |
    H—C—OH
        |
        C=O
        |
    H—C—OH
        |
    H—C—OH
        |
    H—C—OH
        |
    H—C—OH
        |
    H—C—OH
        |
        H
```

19.3 The segment of DNA that has the base sequence T-A-C-G will have the following base sequence in the complementary chain: A-T-G-C.

19.4 The amino acid sequence Glu-Met-Tyr can have one of four different sequences in the messenger chain. These are

G-A-A-A-U-G-U-A-C   G-A-A-A-U-G-U-A-U

G-A-G-A-U-G-U-A-C   G-A-G-A-U-G-U-A-U

■ Answers to Questions to Test Your Reading

19.1 The two types of cells are prokaryote and eukaryotes. Prokaryote are single-celled organisms, such as bacteria. Eukaryotes are all multi-cell organisms, such as plants and animals, as well as some single-celled organisms, such as yeast. The main difference between the two is the presence in eukaryotes of organelles, which are parts of the cell having distinct functions, such as the nucleus.

19.2 The nucleus directs the workings of the cell. It contains the information for the preparation of proteins and also the information that is passed to the next generation. Mitochondria are organelles that produce the energy of the cell from food molecules, such as glucose.

19.3 The four classes of biological molecules are proteins, nucleic acids, carbohydrates, and lipids. Proteins are chain polymers composed of amino acids, which are organic molecules containing an amine group and a carboxylic acid group. Carbohydrates are either simple sugars or else compounds consisting of two or more simple sugar units. They have a carbonyl group and two or more hydroxyl groups. Nucleic acids are polymers composed of nucleotide units. These are either deoxyribose or ribose to which a phosphate unit and a nitrogen-containing base are attached. Lipids are biological substances that are soluble in organic solvents. These include steroids and fats.

19.4 Proteins can function as catalysts and are called enzymes. They also function as molecular transporters. An example is lipoproteins. They also function as structural support as in skin and bone. Or, they function in movement as in muscle tissue.

19.5 Examples of carbohydrates include the sugars, such as glucose (or dextrose) and sucrose. Starch is an example of a glucose polymer, as in cellulose.

19.6 The two types of nucleic acids are deoxyribonucleic acid (DNA) and ribonucleic acid (RNA). DNA is the repository of a cell's genetic information. The RNA is a messenger molecule that forms a template for the synthesis of a given protein molecule.

19.7 The lipids include the following two groups. The steroids are fused-ring compounds that include sex hormones and cholesterol. The fats, or triacylglycerols are molecules that consist of glycerols bonded through ester linkages to three fatty acids. Another type of fat is a phospholipid, which is a constituent of cell membranes.

19.8
$$H_2N-\overset{\overset{\displaystyle CH_2OH}{|}}{\underset{\underset{\displaystyle H}{|}}{C}}-COOH$$

the amino acid is serine

19.9
$$CH_3-\overset{\overset{\displaystyle H}{|}}{\underset{\underset{\displaystyle NH_2}{|}}{C}}-\overset{\overset{\displaystyle H}{|}}{\underset{\underset{\displaystyle H}{|}}{C}}-COOH$$

a β-amino acid

19.10 Ala-Gly

$$\underset{\underset{\displaystyle H}{|}}{\overset{\overset{\displaystyle H}{|}}{N}}-\underset{\underset{\displaystyle H}{|}}{\overset{\overset{\displaystyle CH_3}{|}}{C}}-\overset{\overset{\displaystyle O}{\|}}{C}-\underset{\underset{\displaystyle H}{|}}{\overset{\overset{\displaystyle H}{|}}{N}}-\underset{\underset{\displaystyle H}{|}}{\overset{\overset{\displaystyle H}{|}}{C}}-\overset{\overset{\displaystyle O}{\|}}{C}-OH$$

It differs from Gly-Ala in the location of the CH_3 group.

19.11 Tripeptides of Phe, Ala, and Gly are:

  Phe-Ala-Gly   Ala-Phe-Gly   Gly-Ala-Phe
  Phe-Gly-Ala   Ala-Gly-Phe   Gly-Phe-Ala

19.12 The primary structure of a protein or a peptide refers to the order, or sequence, of the amino acid units in the protein or peptide molecule, and also the locations of any disulfide bonds that might be present.

19.13

[Reaction diagram: Two cysteine-containing amino acid structures with –SH groups + (O) → disulfide-linked structure (–S–S–) + H₂O]

19.14 Digestion is the reverse process of building proteins from amino acids. Enzymes present in digestive fluids catalyze the breaking of the peptide bonds to yield amino acids.

19.15 Although kidney beans are incomplete proteins in that they do not have all the essential amino acids, when combined with other protein foods, such as rice, the combination of foods has all the essential amino acids and is a complete protein diet.

19.16 In globular proteins, the coils or helixes of the proteins fold themselves into compact spherical shapes (or globules) in which the nonpolar side chains of the protein, which are not attracted to water, are inside the globule. The polar groups are on the outside of the globule, where they are attracted to water. The forces responsible for the action are the dispersion forces between the nonpolar portions of the chain on the inside of the globule and also dipole-dipole forces between the polar groups and the water on the outside of the globule.

19.17 When a protein is denatured, it loses its three-dimensional shape through the unfolding and uncoiling of the protein as a result of the breaking of the weak forces, such as hydrogen bonding, that holds the protein in its normal three-dimensional shape. The normal biochemical function is lost. However, if the denatured protein is purified and oxidized, it reverses its shape and activity.

19.18 The term carbohydrate arises from the general molecular formulas of these substances, $C_m(H_2O)_n$. The actual structures of these compounds is very different from this. They are really polyhydroxy ketones or aldehydes, or else a substance that yields such compounds if the carbohydrate hydrolyzes with water.

19.19

glyceraldehyde, an aldose   dihydroxyacetone, a ketose

19.20 A monosaccharide is a simple sugar such as glucose, fructose, ribose, and 2-deoxyribose. An oligosaccharide is a short polymer of two to ten monosaccharide units. An example is sucrose, which contains a glucose and fructose units. A polysaccharide is a polymer containing many monosaccharide units. Starch and cellulose are examples of glucose polymers.

19.21 The final compound obtained in the complete digestion of starch is the monosaccharide glucose. This is a hydrolysis reaction, or a reaction with water.

19.22 Nucleic acids are the molecular carriers of genetic information. The sugar constituent of DNA, the primary carrier, is 2-deoxyribose.

19.23 The four bases contained in DNA are cytosine, thymine, adenine, and guanine. In RNA, the bases are cytosine, uracil, adenine, and guanine.

19.24 ATP is used by organisms as an immediate source of energy. When glucose is broken down in a cell, the energy obtained is stored in the ATP molecule by the following reaction.

$$\text{ADP + phosphate ion + energy} \longrightarrow \text{ATP} + H_2O$$

During biochemical processes that require energy, the reverse process occurs.

19.25 DNA consists of a strand of two complementary polynucleotide chains twisted together into a double helix. The sequence of different bases on the chain is a code representing the sequence of amino acids in a series of protein molecules. During cell division, the coil unravels and duplicates, producing two cells, each with its own DNA double helix.

19.26 The double helix structure is due to complementary base pairing, which is pairing through hydrogen bonding, of certain bases in DNA with others. Certain base pairs, such as adenine and thymine, form multiple hydrogen bonds with each other. One of the polynucleotide strands in a length of DNA is the complement of the other strand.

19.27 RNA is much smaller than DNA. This is because the molecules, such as messenger RNA (mRNA) are complementary copies of a segment of DNA for a particular protein. DNA contains the sequences for many proteins.

19.28 The flow of protein information begins with transcription, in which the information for the sequence of amino acids in a protein is transferred from a DNA chain in the cell nucleus to a messenger RNA molecule. The messenger RNA molecule migrates from the cell nucleus to the cytoplasm, where it serves as a template on which the amino acids are assembled in a process called translation.

19.29 Phenylalanine (Phe) has two possible codons: UUU and UUC.

19.30 The transfer RNA (tRNA) molecules are smaller than mRNA molecules and actually perform the translation of the RNA base code. Each codon, or base triplet, on the mRNA chain links only to a tRNA molecule having the complementary anticodon. Each tRNA molecule in turn carries a particular amino acid.

19.31 A lipid is a biological substance belonging to one of several structurally different classes of substances that are soluble in organic solvents, such as chloroform, $CHCl_3$. Two examples are triacylglycerols (fats) and phospholipids.

19.32 When a fat (triacylglycerol) is heated with a solution of sodium hydroxide (lye), it is hydrolyzed, yielding glycerol and the sodium salts of fatty acids, or soap. The process is called saponification.

19.33 A soap is a long molecule with a nonpolar end and a carboxylate ion end, which is polar. An example is the stearate ion, $CH_3(CH_2)_{16}CO_2^-$. When these ions are in water, they form micelles that absorb the oil and grease into the hydrocarbon (nonpolar) centers while the polar ends are outward toward the water, and can be washed away.

19.34 Phospholipids form bilayers in which the nonpolar ends of the molecules are inside the sheet and the polar ends are outside. They form sheets because the two fatty acid groups are bulky and prevent the formation of more compact micelles. Triacylglycerols are not as bulky and can form the more compact micelles.

Solutions to Practice Problems

19.35

alanine

CH₃CHCOOH
|
NH₂

β-amino acid

CH₂CH₂COOH
|
NH₂

19.37

$$H-N(H)-C(H)(CH_3)-C(=O)-OH \;+\; H-N(H)-C(H)(CH_2C_6H_5)-C(=O)-OH \longrightarrow$$

$$H-N(H)-C(H)(CH_3)-C(=O)-N(H)-C(H)(CH_2C_6H_5)-C(=O)-OH \;+\; H_2O$$

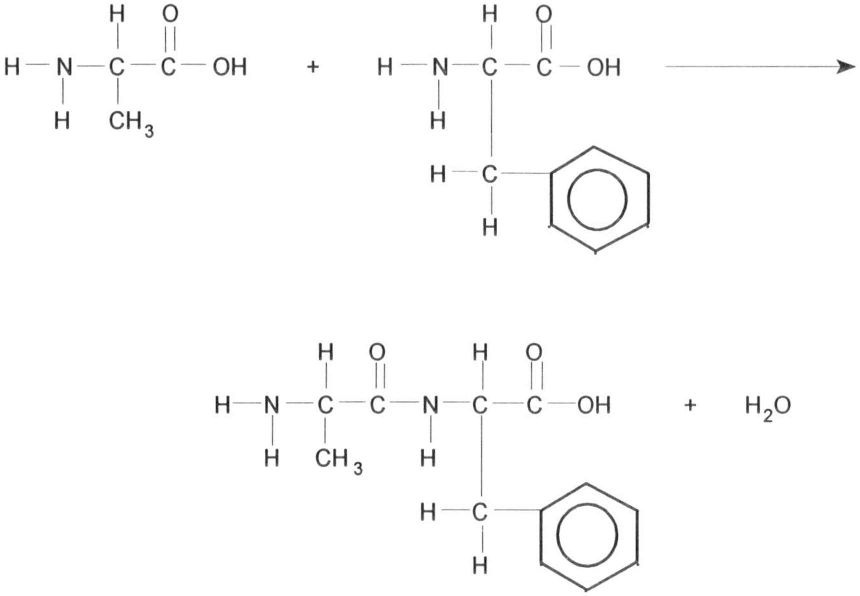

(continued)

H-N(H)(H)-C(H)(CH2-C6H5)-C(=O)-OH + H-N(H)(H)-C(H)(CH3)-C(=O)-OH →

H-N(H)-C(H)(CH2-C6H5)-C(=O)-N(H)-C(H)(CH3)-C(=O)-OH + H2O

The two product molecules have completely different structures.

19.39 Glycine (Gly), alanine (Ala), methionine (Met), and leucine (Leu)

Gly-Ala-Met-Leu	Ala-Gly-Met-Leu	Met-Gly-Ala-Leu	Leu-Met-Gly-Ala
Gly-Ala-Leu-Met	Ala-Gly-Leu-Met	Met-Gly-Leu-Ala	Leu-Met-Ala-Gly
Gly-Met-Ala-Leu	Ala-Leu-Gly-Met	Met-Ala-Gly-Leu	Leu-Gly-Met-Ala
Gly-Met-Leu-Ala	Ala-Leu-Met-Gly	Met-Ala-Leu-Gly	Leu-Gly-Ala-Met
Gly-Leu-Ala-Met	Ala-Met-Leu-Gly	Met-Leu-Ala-Gly	Leu-Ala-Gly-Met
Gly-Leu-Met-Ala	Ala-Met-Gly-Leu	Met-Leu-Gly-Ala	Leu-Ala-Met-Gly

The amine groups are on the left and the carboxylic acid groups are on the right.

19.41 aldoses:

5-carbon:
$$\begin{array}{c} O \\ \| \\ C-H \\ | \\ H-C-OH \\ | \\ H-C-OH \\ | \\ H-C-OH \\ | \\ H-C-OH \\ | \\ H \end{array}$$

6-carbon:
$$\begin{array}{c} O \\ \| \\ C-H \\ | \\ H-C-OH \\ | \\ H-C-OH \\ | \\ H-C-OH \\ | \\ H-C-OH \\ | \\ H-C-OH \\ | \\ H \end{array}$$

19.43 $C_{12}H_{22}O_{11} + H_2O \xrightarrow{H^+} C_6H_{12}O_6 + C_6H_{12}O_6$

      Lactose   Glucose   Galactose

19.45 A-T-G. The base sequence of the complementary chain is T-A-C.

19.47 Ala-Gly-Phe. There are many possible base sequences. One is: GCAGGAUUU. (See Table 19-3 for the various codons)

19.49 For a GCAGGAUUU sequence in m-RNA, the corresponding base sequence in DNA would be CGTCCTAAA.

19.51
$$\begin{array}{c} H \quad\quad\quad O \\ | \quad\quad\quad \| \\ H-C-O-C-(CH_2)_{10}CH_3 \\ | \quad\quad\quad\quad O \\ \quad\quad\quad\quad \| \\ H-C-O-C-(CH_2)_{16}CH_3 \\ | \quad\quad\quad\quad O \\ \quad\quad\quad\quad \| \\ H-C-O-C-(CH_2)_7CH=CH(CH_2)_7CH_3 \\ | \\ H \end{array}$$

19.53

$$\begin{array}{c}
 H O \\
 H-C-O-C-(CH_2)_{10}CH_3 \\
 O \\
 H-C-O-C-(CH_2)_{16}CH_3 \\
\end{array}$$

$$(CH_3)_3\overset{+}{N}-CH_2CH_2-O-\underset{O^-}{\overset{O}{\underset{\|}{P}}}-O-\underset{H}{\overset{H}{C}}-H$$

■ Solutions to Additional Problems

19.55 alanine (Ala), glycine (Gly), and valine (Val). There are six possibilities.

Ala-Gly-Val	Gly-Ala-Val	Val-Gly-Ala
Ala-Val-Gly	Gly-Val-Ala	Val-Ala-Gly

19.57 The disulfide bond forms between the cystine units.

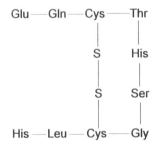

19.59 DNA base sequence TTGCTTCAA would copy to a messenger RNA as AACGAAGUU. This base sequence corresponds to the amino acid sequence Asn-Glu-Val. (See Table 19-3 for the various codons)

Solutions to Practice in Problem Analysis

19.1 In order to write the structural formula of Gly-Ala-Ser, first refer to Table 19.1 and determine the structures of the amino acids glycine (Gly), alanine (Ala), and serine (Ser). Write the structural formula of the three amino acids in the order given, with the amine end on the left and the carboxylic acid end on the right. Next, connect the acid end of glycine to the amine end of alanine, eliminating H_2O in the process, and forming a peptide bond. Then, connect the acid end of alanine to the amine end of serine, eliminating H_2O in the process, and forming a peptide bond.

19.2 To determine the possible base sequences for the portion of DNA that corresponds to Gly-Ala-Ser, refer to Table 19.3 and look up the codons that correspond to each amino acid in the sequence. There are four codons in the table for each amino acid. To construct a base sequence, arbitrarily pick a codon for each amino acid. There are 64 possible base sequences that can be constructed for the amino acid sequence.

Answers to the Practice Exam

1. a   2. b   3. d   4. b   5. c   6. e   7. b   8. b

APPENDIX A. MATHEMATICAL SKILLS

■ Solutions to Exercises

1. a. $5.2 \times 10^9 + 6.3 \times 10^9 = 11.5 \times 10^9 = 1.15 \times 10^{10}$
    b. $3.142 \times 10^6 + 2.8 \times 10^4 = 3.142 \times 10^6 + 0.028 \times 10^6 = 3.170 \times 10^6$
    c. $3.142 \times 10^6 - 2.8 \times 10^4 = 3.142 \times 10^6 - 0.028 \times 10^6 = 3.114 \times 10^6$
    d. $4.1 \times 10^8 + 4 \times 10^7 = 4.1 \times 10^8 + 0.4 \times 10^8 = 4.5 \times 10^8$
    e. $4.1 \times 10^8 - 4 \times 10^7 = 4.1 \times 10^8 - 0.4 \times 10^8 = 3.7 \times 10^8$

2. a. $5.4 \times 10^{-7} \times 1.8 \times 10^8 = (5.4 \times 1.8) \times 10^1 = 9.\underline{7}2 \times 10^1 = 97$
    b. $\dfrac{3.0 \times 10^{-6}}{6.0 \times 10^3} = \dfrac{3.0}{6.0} \times 10^{-9} = 0.50 \times 10^{-9} = 5.0 \times 10^{-10}$
    c. $\dfrac{3.0 \times 10^6}{6.0 \times 10^{-3}} = \dfrac{3.0}{6.0} \times 10^9 = 0.50 \times 10^9 = 5.0 \times 10^8$
    d. $6.1 \times 10^9 \times 2.3 \times 10^2 = (6.1 \times 2.3) \times 10^{11} = 1\underline{4}.0 \times 10^{11} = 1.4 \times 10^{12}$
    e. $\dfrac{2.5 \times 10^4}{5.0 \times 10^6} = \dfrac{2.5}{5.0} \times 10^{-2} = 0.50 \times 10^{-2} = 5.0 \times 10^{-3}$

3. a. $b = \dfrac{V}{T}$ b. $T = \dfrac{PV}{nR}$ c. $T_2 = \dfrac{V_2 T_1}{V_1}$
    d. $\text{volume} = \dfrac{\text{mass}}{\text{density}}$ e. $°F = \dfrac{9}{5} °C + 32$

4. At point C, the volume of the gas is 1 L, and the pressure is 4 atm.

APPENDIX B. USING YOUR CALCULATOR

Solutions to Exercises

1. a. $5.67 \times 10^{-3} + 2.1 \times 10^{-4} = 5.88 \times 10^{-3}$

    b. $3.0 \times 10^7 - 3 \times 10^6 = 2.7 \times 10^7$

    c. $4.3 \times 10^{-2} \times 9 \times 10^{10} = \underline{3}.9 \times 10^9 = 4 \times 10^9$

    d. $\dfrac{7.2 \times 10^{-5}}{3.6 \times 10^{-7}} = 2.0 \times 10^2$

    e. $\dfrac{4.4 \times 10^3 \times 1.7 \times 10^{-2}}{5.1 \times 10^{-3} \times 4.4 \times 10^{-5}} = 3.\underline{3}3 \times 10^8 = 3.3 \times 10^8$

    f. $\dfrac{2.5 \times 10^{-2} + 1.0 \times 10^{-1}}{5.5 \times 10^{-2} - 5.0 \times 10^{-3}} = \dfrac{1.25 \times 10^{-1}}{5.0 \times 10^{-2}} = 2.5$

    g. $1.1 \times 10^2 + 2.0 \times 10^3 \times 1.5 \times 10^{-2} - \dfrac{3.6 \times 10^9}{2.4 \times 10^7} = -10.$